Photovoltaic System Design

Procedures, Tools and Applications

T0199514

Suneel Deambi

CRC Press
Taylor & Francis Group
Boca Raton London New York

CRC Press is an imprint of the
Taylor & Francis Group, an **informa** business

CRC Press
Taylor & Francis Group
6000 Broken Sound Parkway NW, Suite 300
Boca Raton, FL 33487-2742

First issued in paperback 2020

© 2016 by Taylor & Francis Group, LLC
CRC Press is an imprint of Taylor & Francis Group, an Informa business

No claim to original U.S. Government works

ISBN 13: 978-0-367-57463-5 (pbk)
ISBN 13: 978-1-4822-5980-3 (hbk)

Library of Congress Cataloging-in-Publication Data

Names: Deambi, Suneel, author.
Title: Photovoltaic system design : procedures, tools and applications / Suneel Deambi.
Description: Boca Raton : Taylor & Francis, 2016. | Includes bibliographical references and index.
Identifiers: LCCN 2015045365 | ISBN 9781482259803 (hardcover : alk. paper)
Subjects: LCSH: Photovoltaic power systems--Design and construction.
Classification: LCC TK1087 .D434 2016 | DDC 621.31/244--dc23
LC record available at http://lccn.loc.gov/2015045365

**Visit the Taylor & Francis Web site at
http://www.taylorandfrancis.com**

**and the CRC Press Web site at
http://www.crcpress.com**

Contents

List of Figures

List of Tables

Preface

The world is currently assuming the form of a close-knit communication platform with or without access to electricity in several parts. Mobile telephony is fast spreading its wings far and wide, with those hooked on to it finding one channel or another to charge a mobile phone, for instance. Incidentally, there are several villages in India where electricity is unavailable, but happy phone users commute to adjacent areas to charge their phones. This raises the pertinent question of whether one can continue to live without electricity but not without a mobile phone. Of course, securing electricity access far outweighs other considerations, but it is simply unavoidable to align with modern-day transformations. This brings into focus the role of clean, safe and reliable green technologies such as solar photovoltaics (PV). Today, the role of PV technology is far more etched in several ways than ever before. Rooftops are fast becoming the agents of new change in terms of accommodating solar PV panels along with a mushrooming number of direct-to-home (DTH) dishes for viewing TV programmes. The list of end users taking an early recourse to PV system adoption grows by the day. Now, 'solar' is no longer an alien word, but the real issue hovers around its long-term reliable use.

It is worth mentioning here the repeatedly talked about number of ill-performing PV systems for various end-use applications, including lighting, water pumping, battery charging (for multiple uses) and, most importantly, power generation. We should also keep in view the fact that non-solar appliances malfunction at times. So solar is no exception on this front. The fact is that PV systems have now become more reliable than ever before, even though it still makes sense to keep monitoring their performance under actual field operating conditions.

Proper selection of a solar-worthy site from all possible considerations is fast becoming a norm rather than a choosy affair. A majority of PV systems now rely on the use of simulation software, for example, to determine power generation. Site resource assessment is now an important activity preceding the actual system sizing procedure. Several PV simulation softwares are currently available in the marketplace with different degrees of actual utilization across the world. However, the underlying logic is to first gain familiarity of the key factors prior to undertaking a system simulation exercise. The moot question is whether manual system sizing procedures are better than simulated procedures any day. Well, the latter techniques offer a plethora of benefits due to the embedded nature of checks and balances within the software.

A larger issue is the capability to utilize the simulation software available in the market today. In a country like India, obtaining an individual licence to softwares is mostly out of the question due to their high costs. However, there are several softwares with identical purposes available for free download most of the time. This brings into the realm of immediate discussion the curiosity of getting the most accurate results while dealing with PV systems under actual field operating conditions. Several parametric considerations delve deeply into the relevant scheme of things so as to suggest that a system simulation activity is not a cakewalk. It needs to be performed with the best of system design and engineering capabilities so as to romp home with a good enough harvest of the expected energy generation. The real need is to devise a true synergy amongst all the factored-in values of actual importance in the system size estimation and the accompanying values of the expected system performance. Capacity building, vis-à-vis these modern-day tools and procedures, is an important activity of the whole charter, failing which not much can be accomplished on this important front.

According to the market analysis made by GTM Research, global solar PV installations for 2015 touched a high of 59 GW. This marks an increase of around 34% over the total installed PV capacity achieved in 2014. Further, there is an expected cumulative realisation of worldwide PV capacity of about 321 GW by the end of 2016. Geographically, the largest markets, yet again, are China, Japan and the United States as per the available market estimates of a well-known PV market information company with diverse interest in several core areas including energy (i.e. IHS). Grid-connected PV energy storage installations grew in 2015, though not appreciably. Likewise, PV systems up to 100 kWp accounted for as much as 20–30% of global installations. Monocrystalline silicon cell technology is gradually inching forward to retrieve its lost ground to the currently-in-vogue polycrystalline technology. India has recently upscaled its PV installation target to around 1,00,000 MW by 2022, which seems to be a gigantic target by any means. The sunrise technology sector in India is set to gain expanded horizons of use across the diverse sectors of energy economy. This really signals a heightened impetus for spreading awareness about PV simulation tools and procedures in no uncertain terms.

This book lays down a preparatory framework on the diverse aspects of PV technology, applications and programmes from a variety of end-use considerations. It is expected to serve as a basis for realizing expected gains from PV systems while taking recourse to simulation activity. There are eight chapters in the book with clearly spelt-out objectives and expected outcomes. Chapter 1 touches upon the role of renewable energy technologies in a holistic energy scenario in terms of several projected demand and supply scenarios and underlying solutions. Chapter 2 makes a clear categorization of off-grid and on-grid PV applications while drawing upon the relative advantages and limitations. Chapter 3 highlights the significant promise of solar radiation availability on Earth in terms of vital dependencies on several processes, phenomena and cyclic variations of several elemental considerations. Chapter 4 takes up the issues related to on-site-specific considerations. It is crucial to gain a clear understanding of the issues which may impact the operation and smooth running of PV facilities in one way or another. Chapter 5 involves a basic treatment of the system design considerations from a variety of parametric aspects and site-specific prevalence of the important issues. Chapter 6 introduces PV system sizing procedures via the modern use of simulation software. The specific objective is to draw out clear advantages of basing a system design on the resultant outcome of a simulation activity from a variety of end-use considerations. Each of these softwares holds out relative merits and demerits in terms of its applicability to a given field situation. Chapter 7 presents an analysis of actual PV power plant sites when designed via the use of simulation software. It goes on to determine the weak links in a PV system design-cum-engineering chain, for instance, in terms of the assessed number of failures as attributed to that from real-time monitoring of the systems. Finally, Chapter 8 brings out the clear importance of capacity-making initiatives vis-à-vis the available range of PV simulation software, tools and, most importantly, procedures at large. It takes up a few case-specific examples of education, training and awareness generation activities of a selective few organizations from a practical perspective.

Overall, this book attempts to familiarize interested readers with solar PV technology and its key attributes, end-use applications, system design requirements, influence on climatic and site-specific parameters, utilization of simulation procedures and expected performance levels in an easy-to-understand manner.

Acknowledgements

I am very grateful for the support of various stakeholders who are involved in the overall development of the solar PV programme in India. Without their help, it would have been quite difficult to put this book together. It is not possible for me to acknowledge each source of information individually. However, I extend my sincere thanks to all of them.

I owe my profound sense of gratitude to CRC Press, who motivated me and Dr. Om. P. Nangia (ex-BHEL) to take up this topic from an Indian perspective. Although Dr. Nangia could not continue being part of this project, I thank him for his help at the initial stages of the preparation of this book.

Finally, I dedicate this book to the cherished memory of my late father, Sh. Brij Nath Deambi. My affectionate thanks also go to my wife, Neeru Bhat Deambi, and daughter, Tammana Deambi, for their patience and support throughout the stages of the book preparation.

About the Author

Suneel Deambi, PhD, is a solar energy specialist with practical experience of about 25 years in the renewable energy (RE) sector. His active areas of interest are RE policy, planning, technology, financing, programme implementation, performance evaluation and capacity building initiatives. He is a prolific writer on energy–environment issues with three books to his credit plus a large number of published articles/features/reviews etc. in the leading media of information dissemination.

Chapter 1

Role of Renewable Energy Technologies in the Overall Energy Scenario at a Global Level

1.1 Introduction

As per the International Energy Agency, around 1.3 billion people do not have access to electricity, mainly people living in the rural regions of Asia and sub-Saharan Africa. Incidentally, just about 5% of sub-Saharan Africa has access to energy. Electricity consumption in this region is only as much as in the state of New York. Importantly enough, the challenge of energy accessibility must be viewed in terms of availability as well as affordability for individuals and communities. A good thing about renewables is that renewable energies (REs) such as wind, solar or biomass are available in small units, that is as solar cells or wind turbines. This makes their construction and maintenance fairly straightforward. Over time, the costs for solar, wind and renewable biomass have dropped down significantly. Importantly, some countries have already made remarkable inroads in distributing RE technologies. Take for instance a small country like Bangladesh where more than 80,000 solar home systems are being installed month after month in rural off-grid areas. The RE sector in Bangladesh has employed as many as 200,000 personnel.

According to the World Resource Institute's (WRI) report, between now and 2050, countries in the Global South need an estimated $531 billion per year of additional investment in energy distribution, grid and storage systems. This is expected to limit the global temperature rise by 2°C above the preindustrial levels. It is now widely believed that sound national and regional policies that spur the energy transition and development of renewables are key to attaining progress. At a global level, feed-in-tariffs have proven to be an effective means to quickly increase the generation of renewable-derived electricity.

Feed-in-tariffs or simply FITs accord the small-scale power producers an option of being paid for feeding their electricity into a regional or national electricity grid. These payments have been showcased to be more effective than the other RE policies such as direct investment subsidies in terms of pricing, innovation and job growth. FITs, when made adoptable to the local context of a country or region, can enhance overall energy production and boost economic development besides improving access to energy. In addition, the decentralized approach offers a sound opportunity for community-owned energy production.

However, creating a reliable environment for investments in RE is one of the biggest challenges for the industry. As the availability of domestic finance is often a key barrier to rolling out FITs in domestic countries, the newly established Green Climate Fund of the United Nations Framework

Convention on Climate Change aims to provide an effective solution. There is already some global experience in implementing FITs using international climate financing. Seemingly, little or no pledges to the Green Climate Fund have been made so far. However, it represents a promising attempt to address the challenges that RE investors face in developing countries. Approaches such as FITs are real opportunities that encourage the development of RE in the Global South.

1.2 Global Energy-Use Status

There are several key drivers of global energy demand such as increasing population. The global population in 2011 was around 7 billion as against a projected number of 8 billion by 2030. The growing economies of the world are BRICS and N11 countries in a broader context. Approximately 1.3 billion people do not have access to electricity and 2.6 billion people even today rely more on wood, dung, coal and other traditional fuels in their homes. In sum, energy demand is increasing at a rate of around 2% per annum. Following are the few core objectives as enshrined under the international energy policy:

- To double the share of RE in the global energy mix
- To ensure universal access to modern energy services
- To double the global rate of energy efficiency in buildings, agriculture and transportation sectors, etc.
- To contribute to food, air and water security

1.2.1 Role of Private Sector

Any day the private sector is expected to contribute its capacity, knowledge, experience and targeted funds, besides the flexibility and much needed efficiency, in implementing the earlier measures. Amongst others, these private entities include major pension funds, mutual funds, sovereign wealth funds, private corporations, development banks and other investors. It is worth mentioning here that a limited use of recent technological advancements has been made so far in energy planning and monitoring activities. Also decision makers at times do not capture the complexity and spatial patterns of the socioeconomic systems.

The following section takes a close look at the resource mix from several key considerations.

1.3 Resource Mix

Energy has always been the most important factor and the prime mover of global economy. All energy sources play significant roles in achieving a reliable access to energy for the masses in an efficient manner. Today, over 20% of the global population lives without access to electricity and an equivalent share is estimated to have only an intermittent access in the developing world. Energy is the key driver of our economy in real terms. It is thus of special relevance to continue producing from all the possible sources available to us on earth. This is called 'effective energy resource mix'. To meet the growing energy needs, besides preserving the environment, societies have to put in place a varied energy mix over the long term, with reduced greenhouse gas (GHG) emissions. Combined with energy savings, this technological and political development is also the key to a more secure energy supply. This ongoing development towards the supply and demand of goods and services is acknowledged by the societies. Against a backdrop of globalization, this process extends to emerging and developing countries. In China, India and Brazil, a new middle class is gradually emerging;

they expect the modern conveniences that are already wide-spread in Europe and North America (such as household appliances and personal vehicles). Take, for example, the often-talked-about fact that India has the largest chunk of middle class families in the world.

To ensure sustainable development, global energy mix needs to be diversified in the twenty-first century. This economic growth goes hand in hand with growing energy needs on a global scale, most of which are met by fossil fuels. However, these energy sources are getting scarce, and it is not possible to increase oil, gas or coal production capacity indefinitely. Furthermore, using fossil fuels creates significant GHG emissions. According to the Intergovernmental Panel on Climate Change (IPCC), 75% of the CO_2 emitted between 1981 and 2001 was from the burning of oil, gas and coal; these GHGs play a key role in global warming today.

Therefore, to ensure human development, besides pre-serving the environment, we need to diversify our global energy mix in the twenty-first century. This will involve using alternative energy sources with few GHG emissions. Adopting the energy and climate change package of Europe is a part of this approach. Energy and climate change package is an action plan whereby the EU has committed to meet 20% of its energy needs through RE by 2020, as compared to 8.5% in 2008. Importantly enough, as we diversify, we also need to save available energy and optimize its use. By using less energy-intensive appliances or vehicles, we can help reduce energy demand. These appliances use energy more efficiently, thus increasing national energy intensity (the ratio between energy consumption and the wealth provided with this energy, represented by a country's GDP). However, global energy needs are so high that this is not enough to reduce energy demand on its own; it will only slow down its growth.

1.3.1 Various Policies to Match Energy Supply and Demand

As the world's energy needs are expanding, new challenges are taking shape. In the next few years, humanity will have to meet a dual energy challenge. They are briefly indicated here:

■ In the medium term (within 15–20 years), oil supply will begin to taper off while demand continues to grow.
■ In the long term, after 2050, fossil fuels will diminish, starting with oil, followed by gas about 20 years later. At this stage, we have enough coal to last for about 20 years.

To meet these crucial challenges, we need to make significant investments. Consequently, political and economic stakeholders have to make choices, including diversifying and decarbonizing the energy mix (developing energy sources with a low carbon footprint). They have developed a number of ways to match energy supply to demand. Some methods are compatible, and even complementary, and all have their own advantages and disadvantages.

1.3.2 Country-Specific Examples in Brief

The United States and Australia consume a lot of domestic energy resources. The United States is largely dependent on oil and it is trying to reduce its energy consumption and encourage RE development so as to preserve its energy independence. On the whole, this approach will allow people to retain their lifestyle for another few decades, sparing them the upheaval associated with changes in the way society operates. The energy and climate change package adopted by Europe in the late 2008 and the measures set forth in the Kyoto Protocol also aim to control demand.

By anticipating shortages, these policies can soften their effects, for example reducing energy consumption prolongs the life-span of fossil fuels. This gives governments the time to develop alternative resources to fully replace oil in the energy mix over the longer term. Moreover, these measures help reduce GHG emissions. However, they require citizens to agree to take part in efforts to save energy by changing their lifestyles in terms of transportation, housing and every-day life. Whatever their policies, most rich and emerging countries encourage the development of alternative energies (nuclear and renewable).

Apart from their environmental qualities (no GHG emissions), these types of energy choices help delay the depletion of fossil fuels. However, current research is not yet at a stage where it can replace oil entirely. In future, nuclear energy will be one of the more promising alternative energy sources. Nuclear fusion could provide an almost inexhaustible energy supply. But these technologies are very expensive and very difficult to control and will not be part of the energy mix before 2050.

1.3.3 Security of Energy Supply

Oil is still regarded with a sufficient degree of reliability as the primary source of fuel across the world. During the twentieth century, oil became a strategic resource, indispensable to running our economies and societies. However, oil and gas supply is subject to geopolitical issues – a tense international situation can block energy feedstock trading. The most recent oil shortage in the Western countries dates back to the first oil crisis of the 1970s. Thus, in the past few decades, industrialized countries have become used to having lots of cheap oil and gas. Alternative energy seemed expensive in comparison. With the globalization of the world economy, emerging and developing countries have also become highly dependent on the energy supplied by the hydrocarbon-producing countries, which belong to two main groups (i.e. organisation of petroleum exporting countries and International energy agency). The Middle East is rich in oil and gas deposits. The Organization of Petroleum Exporting Countries active in this region plays a pivotal role in managing production worldwide apart from stabilizing crude oil prices. Furthermore, Russia and countries bordering the Caspian Sea (Kazakhstan and Turkmenistan) supply Europe and Eastern Asia. The construction of the Baku–Tbilisi–Ceyhan (BTC) pipeline linking the Caspian Sea to the Mediterranean has enabled the states in that region to sell their oil directly to European countries.

Oil and gas are also found in South America, Canada and the Niger Delta in Africa. These hydrocarbons are mostly useful for the United States and China too. Also, some North African oil and gas is sent to European countries. Hydroelectric dams can cause tensions if located on large rivers. The geopolitics of other resources is less problematic; for example coal is widely distributed worldwide and it is produced by consumers themselves. As for uranium reserves, they are partly held by rich countries (e.g. Australia, the United States and Canada). However, over 80% of the global energy mix comes from fossil fuels. Therefore, any serious geopolitical crisis in a large oil-producing country could affect the whole planet. Faced with this risk, consumer countries have formulated a number of solutions. These are briefly mentioned in the following:

■ To bring down their dependence on oil by developing alternative energy resources so that they have a more varied energy mix (e.g. developing countries have significant solar, wind, geothermal and hydroelectric resources, which they can develop with the help of rich countries).
■ To adopt maximum energy-efficiency measures. New-generation LED lamps, getting cheaper by the day, can lead to significant energy savings.
■ To secure their energy supply by entering into agreements with certain suppliers.

1.3.4 Share of Fossil Fuels

Fossil fuels – primarily coal, oil and natural gas – account for nearly 80% of the world's primary energy consumption, even though they emit harmful substances. Because of the possibility of inter-fuel substitution in end-use applications, the optimal long-term energy supply requirements of any nation necessitate examination of all energy resources. The GHG effect caused by emissions of pollutant gases from the fossil fuels into the atmosphere result in global warming and cause climate changes. These harmful consequences of the pollutant gases are essentially driving the need for the development of clean and low carbon technologies for power generation; for example, the infinite light energy from the sun or wind energy sustains life on Earth in an eco-friendly manner.

1.3.5 Role of Low-Carbon Technologies

Production of energy by carbon-free technologies will most likely pave the way in fulfilling the initiative of sustainable energy for all. Using low-emission RE technologies and, to an extent, nuclear energy for electricity generation will be a more practical means of overcoming the energy deficit. Solar, wind and biomass technologies which are free of any harmful emissions already have provided a viable solution for delivering energy services in an innovative manner. Compared to the conventional energy systems based on fossil fuels, extensive use of RE systems offer broader achievements such as better health, creation of new avenues for employment and economic development at an enhanced pace. Take for instance the replacement of kerosene oil lamps by solar lanterns; this offers sound health benefits besides offering a good-quality cool white light. At the other end, the fast evolving arena of megawatt-scale photovoltaic (PV) power plants is turning out to be a game changer

Figure 1.1 Global presence of green energy areas.

of sorts by opening up new avenues of job in this multi-disciplinary area.

REmap 2030 of International Renewable Energy Agency (IRENA) invites the world community to forge a new energy future most appropriate to their circumstances, informed by the most comprehensive and transparent data available. Figure 1.1 shows the presence of green energy areas worldwide.

The global RE share can reach and exceed 30% by 2030. Effective technologies are already available today to achieve this objective. Renewable growth needs to take place across all the four critical sectors of energy use: buildings, transport, industry and electricity. Global electricity consumption will continue to grow faster than the total final energy consumption (TFEC) to around 25% of the TFEC in 2030.

1.3.6 *Environmental Advantage*

The large-scale use of renewable and sustainable technologies will benefit human beings and provide environmental protection from floods, draughts and rising sea levels. This will also pave way for rapid deployment and diversity of fuel supply. On account of drastic increase in global power demand and with dwindling fossil fuels reserves, both energy-starved and energy-rich nations have adopted alternative energy sources, available in abundance, in a mission-driven mode. It has, therefore, become essential to increasingly rely on alternate energy resources so as to achieve energy independence and security. The sun is the only sustainable energy source, available freely in abundance with the potential for power generation, is highly safe, is cost effective (availability of wind and biomass is limited), fulfils the long-term energy needs and does not cause climate change. Research and development work has focused on solar and wind technologies for the past 30–40 years worldwide in various laboratories and research institutions; a new imperative for carbon reduction has given these technologies a push. Nuclear technology, though expensive, does not emit CO_2 and thus meets power-generation requirements; however, it suffers from the hazards of radioactive leakage due to unsafe handling or any freak accident that might occur at the installation site during operation and maintenance (O&M) activity at the plant. Solar energy with all its inherent benefits is in high demand and plays a sizeable role in the electricity mix as a preferred source for clean power worldwide.

1.3.7 Energy as an Enabler

The per capita power consumption has become the yardstick to measure the developmental growth and the standard of living for any country, as proven by the relationship between GDP per capita and electricity consumption worldwide (International Energy Agency Report [IEA], Worldwide trends in energy use and efficiency, 2005). Maintaining today's world average energy use per capita is most probably the only thing that can be accomplished, which requires sacrifices in terms of cost, quality and reliability of the energy. Consequently, 100 GJ/ capita/year as a general target (70 kWh/capita/day) is considered highly adequate.

The world's average target in 2008 was 74 GJ/capita/year (52 kWh/capita/day). Energy has a resultant impact on virtually every activity carried out on planet Earth. Following is a brief point-specific treatment of inter-linkages possible on account of energy use in various sectors/sub-sectors:

Energy and Water

- Minimizing the use of water in energy systems (in all steps of the production cycle)
- Maximizing sustainable energy access in all water and sanitation systems

Energy and Health

- Securing sustainable energy for healthcare facilities worldwide
- Eliminating all premature deaths due to air pollution caused by cooking and heating

Energy and Education

- Securing sustainable energy for schools worldwide

Energy and Gender

- Minimizing all risks that women face due to energy-related activities including collecting energy resources, cooking, heating, lighting, etc.

Energy and Food Security

- Reducing the intensity of fossil fuel use in food systems and increasing access to modern energy services while meeting individual and national food requirements

Energy and Environment

- Minimizing all the discharges of contaminants from energy systems to land, atmosphere and water bodies
- Minimizing the rate of deforestation attributed to energy use
- Minimizing GHG emissions from energy systems

Energy and Industrialization

- Reducing industrial energy intensity
- Increasing the use of RE in manufacturing processes
- Providing access to reliable energy services to support structural change and industrialization

1.3.8 Brief Analysis

The aforementioned illustrative examples are possible energy cross-cutting targets that would need further assessment and could be associated with specific quantitative values or indicators within a specific time frame. RE plays a dominant role in contributing to the energy pool. With the alarming depletion of fossil fuel levels coupled with an ever-increasing threat of global warming faced by both the developed and developing world, there is an undeniable need to develop efficient devices so as to convert renewable resources into useful power generators. RE has started to contribute its share towards the global electricity market. RE, excluding large hydro projects, accounted for as much as 43.6% of the new generating capacity installed worldwide in 2013, thus raising its share towards global electricity generation from 7.8% in 2012 to almost a percentage point higher, that is, 8.5%. If this capacity had not been present, world energy–related CO_2 emissions may have been an estimated 1.2 Gt higher in 2013. This means adding about 12% to the 2020 projected emissions gap that needs to be closed to remain within the 2°C global temperature rise.

1.3.9 Assessment of the Available Energy Technologies

In a report by Heinberg*, 9 criteria have been used to assess the potential future of 18 energy sources presently available on the market in a more or less developed form. These criteria can be grouped into six basic categories:

1. Direct monetary cost
2. Environmental impact
3. Renewability
4. Potential scale of contribution
5. Reliability
6. Energy return on energy investment (EROEI)

1.3.9.1 Importance of 'Net Energy'

EROEI, also called 'net energy', is an important criterion. It is ideally seen as a key figure to understand world energy system. The EROEI of U.S. oil was 100:1 in 1930. It fell to 30:1 in 1970 and is currently less than 20:1. According to Heinberg, the high EROEI that oil formerly enjoyed was directly responsible for the development of the energy guzzling economy that we have today. The sense of his argument lies in his assertion that it is very unlikely that we will find a new energy resource with such a high EROEI any time soon. Even though the reserves of oil and natural gas are still significant, the EROEIs of those resources will most probably

---

* Searching for a miracle/net energy limits and the fate of an industrial society, by Richard Heinberg: joint initiative of International Forum in collaboration with the Post Carbon Institute.

continue their steep decrease. This is also the case for coal but to a lesser degree though, since today coal still has an EROEI of around 65:1. Heinberg is not in agreement that coal is available for a few hundreds of years. Rather he predicts the world coal may peak around 2025 accompanied by a steep decline in its EROEI after 2040. The minimum EROEI necessary to sustain a modern industrial society is considered to be 10:1. Carbon capture and storage (CCS) will make the EROEI of coal decline even faster and, for this reason essentially, Heinberg does not see coal with CCS as a sustainable solution. The report concludes that even without taking climate change and other environmental issues into account, we will be forced to shift towards a non-fossil-fuel economy in the coming decades. The potential technologies which would help build a new energy economy are discussed in brief in the following sections.

1.3.9.2 Nuclear Energy

Nuclear energy has many drawbacks – uranium is non-renewable, the initial investments are huge, the environmental impact of the fuel cycle is high and nuclear power plants require a great deal of water. Hydroelectric energy is either on too small a scale and thus does not add up or too large a scale with local environmental and social impacts so prominent that are in most cases too high to be acceptable. Passive solar energy is certainly a valuable concept but too limited in scale to contribute significantly to the world's energy needs. Biomass, biodiesel and ethanol have an EROEI below 5:1.

1.3.9.3 Other Energy Sources

Wind energy, solar photovoltaic (SPV) energy, concentrated solar power (CSP), wave energy and tidal energy are highly potential, but even the potential of this 'energy mix of the future' is limited. PV has drawbacks in its relatively high cost and relatively low EROEI, and the potential of tidal energy is limited to a few regions of the world. Wave energy will need more research before we know its true potential. So Heinberg's main contender is wind energy and concentrated solar power (CSP), which will have to make up the largest share in any viable future energy mix. Offshore wind power development is expected to play a major role alongside the already onshore wind energy realization in the foreseeable future.

1.4 Existing Investment Climate for Renewable Energy Technologies

New investment in RE excluding large hydroelectric projects has slipped from 14% (of the investment made in 2012) in 2013 to $214 billion, but even this disguised one major positive development. One of the two main reasons for this fall in 2013 was a reduction in the costs of PV; even as the dollar investment in solar went down, the number of gigawatts (GW) of PV systems added went up. There were setbacks to investment in many important geographical areas, including China (down 6% at $56 billion), the United States (down 10% at $36 billion) and Europe (down 44% at $48 billion). The biggest exception to such a downward trend was Japan, where investment excluding research and development soared 80% at $29 billion. Renewable solar energy uptake and direct replacement of fossil fuel in the four end-use sectors (buildings, transport, industry and power) are essentially needed in order to reach a doubling of its share. If the REmap options are deployed, the total share of modern RE in 2030 would reach 38% in buildings, 17% in transport, 26% in industry and 44% in power sectors. Around 40% of the total RE potential in 2030 is in power generation, with 60% attributed to the other three end-use sectors.

1.4.1 Introducing Solar Energy Technologies

SPV and solar thermal CSP technology are the two most important routes to harness solar energy. Historically, the selenium-based cell invented in 1885 by Charles Fritt triggered the curiosity of scientists that resulted in the invention of silicon-based solar cells in 1954 at the Bell Laboratories (United States). In an SPV system, sunlight when falls on a solar cell gets absorbed and is converted into direct current. SPV technology is considered more mature compared to CSP in terms of ease of operation for power generation (with no long gestation period) and its utilization in a stand-alone mode and grid-tied applications. Besides, SPV technology has several other advantages such as being modular in nature with no moving parts, no noise, nominal maintenance requirement and being environmentally benign. Advanced R&D has focused on reducing input material requirements and automation in manufacturing in tandem with rapid industrialization and has led to a portfolio of the solar technologies options at different levels of maturity. A wide variety of commercialized solar PV technologies are based on crystalline (mono/multi) silicon, concentrating PV, thin-film technologies (a-Si, Cd Te, Cu, In, Ga, Se_2) with conversion efficiencies ranging between 6% and 30%. Solar power has become affordable now, since PV module costs are falling and the technology is proving highly stable, reliable and adaptable; it is a strong contender in the overall energy mix.

1.4.1.1 Dye-Sensitized Solar Cells

Dye-sensitized solar cells (DSSC) made from organic materials such as dyes are able to trap photons into synthesized electrons that can harvest high levels of photon energy. This technology is still in laboratory scale development and would take some time to scale up to commercial production levels at comparable efficiencies. Several other variants of SPV technologies which are currently being categorized as fourth-generation cells are available.

1.4.2 Concentrated Solar Power Technology

In the various versions of CSP technologies, the conversion of solar energy into heat energy, the incident solar radiation is collected and concentrated by solar collectors or mirrors on a thermal receiver where the thermal energy is used to heat the heat transfer fluid (HTF) such as oil, air or water/steam, depending on the plant design. The HTF is either used directly or is utilized to generate steam or hot gases, which in turn are then used to operate a heat engine. The CSP technology variants are parabolic trough, compact linear Fresnel reflector (or single-axis sun-tracking mirrors) and solar dish and heliostat/power tower (or dual-axis sun-tracking mirrors). The solar parabolic trough has been largely commercialized amongst all CSP technologies. Solar power technology uses a power tower design, which generates power from sunlight by focusing energy onto a tower-mounted central heat exchanger or a receiver. Highly efficient hot molten salt (combination of sodium and potassium nitrate with melting temperature of 46°C) is used for thermal storage of energy.

1.4.3 Grid Parity

Grid parity is the stage at which the levelized cost of electricity generation using commercial solar technologies, becomes competitive to the conventional power generation at peak load levels. Whereas in distributed systems, it is lower than the conventional cost. Given that prices of SPV are falling against the backdrop of rising fossil fuel prices (mainly due to their depleting reserves and enhanced cost of extraction and transportation, etc.), it makes sense for electric utilities all over the world to introduce the promising and cost-competitive solar technology in their generation schemes.

1.4.3.1 Taking the Challenge Head-On

The Sun has for long been recognized as a primal source of all energy on Earth. Advanced R&D needs to be carried out simultaneously in the solar field to achieve higher solar to electric conversion efficiencies at lower costs so as to make them competitive with the conventional sources of power generation. The mature SPV technology systems and devices have an excellent track record worldwide for more than three decades now.

1.4.3.2 Fast-Declining SPV Price Regime

As per the current analysis and prediction, the ongoing trend of falling PV prices will push the total solar installations beyond wind within the next 7 years or so. By 2021, global PV capacity is expected to hit 715.8 GW as against wind's capacity, that is 697.3 MW. Technology costs are a second big reason for the latest fall in investment. PV module prices bottomed out in early 2013 as the industry's severe over-capacity eased and balance-of-plant costs for PV systems continued to fall. In addition, there was a shift in the global mix of PV installations in 2013, with a lower share of relatively high cost per megawatt installed in residential systems and a higher share of relatively low cost per megawatt installed in utility-scale systems (particularly in China). Although PV capacity installed was up from 31 GW in 2012 to 39 GW in 2013, dollar investment in solar capacity was down by around 23% at $104 billion. Utility investment in solar as a wholesale generating resource at the distribution level will also provide new market opportunities besides creating jobs. Figure 1.2* shows the global new investments in RE by asset class for the period 2004–2013.

---

* Global Trends in Renewable Energy Investments 2013: Fraunhofer School with UNEP Collaborating Centre for Climate and basic energy along with Bloomberg Energy Finance.

Growth:

63% 54% 47% 17% −2% 35% 23% −11% −14%

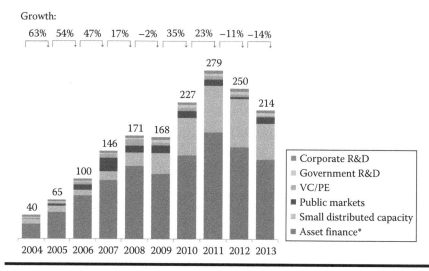

- ■ Corporate R&D
- ▨ Government R&D
- ▨ VC/PE
- ■ Public markets
- ▨ Small distributed capacity
- ■ Asset finance*

Figure 1.2 Global new investment in renewable energy by asset class, 2004–2013. *Asset finance volume adjusts for reinvested equity. Total values include estimates for undisclosed deals. (From UNEP, Bloomberg New Energy Finance, London, U.K [1].)

1.5 Projected Worldwide Goals for the Development and Deployment of Solar Technology

Electrical power touches all aspects of our lives – at home, in transit or at the workplace. Global demand for electricity, being the most essential energy source for economic activities, is set to continue to grow faster than any other form of energy in the coming decades. There is a limited availability of petroleum fuel in the world. Nuclear power plants are considered to be a better option to meet the growing energy demands. However, stringent safety standards in installation and operation of such plants have made this option less attractive. In such a scenario, exploring other options (renewable energy) for power generation becomes much more important. Globally, RE is assuming significant importance for its development. Rapid strides are being made globally by adapting RE technologies. REmap 2030 indicates a pathway for doubling the share of sustainable renewable energy in the world's TFEC. The policies that are currently in place and under consideration would shift the world's renewable share from 18% to 21% in 2030. Doubling the rate of energy efficiency and providing universal access to modern energy services via renewable would raise the share of RE to as much as 36%.

It is a common knowledge that all major RE technologies owe their origin to the Sun. The bounty of the Sun is truly inexhaustible, renewable and free. Solar energy is the undisputed energy of the future. In the post-fossil fuel world, solar energy will encompass all aspects of our life. Increased use of solar energy is a central component of our strategy to bring about a strategic shift from our current reliance on fossil fuels to a pattern of sustainable growth based on renewable and clean sources of energy.

1.5.1 Solar Photovoltaic Leading from the Front

The solar industry was initially nurtured in Germany, Japan and the United States, and then gained strength in countries such as Italy, where government support designed to boost demand helped PV manufacturers increase capacity, reduce costs and advance their technologies. These subsidies helped spur demand that outpaced supply, which brought about shortages that underwrote bumper profits for the sector until 2008. The focus during this period was on developing better cell and module technologies; many Silicon Valley–based venture-capital firms entered the space around this time, often by investing in companies in thin-film solar-cell manufacturing. Valuations for some of the more promising solar-cell start-ups at that time exceeded $1 billion.

The price for installing PV systems in residents fell from more than $100 per watt peak (Wp) in 1975 to $8 per Wp by the end of 2007; although from 2005 to 2008, prices declined at a comparatively modest rate of 4% per year. German subsidies drove value creation, with the lion's share of the value going to polysilicon, cell and module-manufacturing companies in countries that are part of the Organization for Economic Cooperation and Development (OECD). Encouraged by the growth of the industry, other countries – including France, Canada, South Korea, Australia, South Africa, India and China – began to offer support pro-grammes to foster the development of solar sectors within their borders.

Chinese manufacturers began to build a solar-manufacturing sector targeting foreign countries where demand was driven by subsidies, particularly Germany. Armed with inexpensive labour and equipment, Chinese players trig-gered a race to expand capacity that drove PV prices down by 40% per year; prices fell from more than $4 per Wp in 2008 to about $1 per Wp in January 2012. It is estimated that balance-of-system (BOS) costs declined by about 16% per year in this period, from about $4 per Wp in 2008 to approximately $2 per Wp in 2012 (these are more difficult to track, in part because BOS costs vary more than the module costs).

The cost curve flattened for many upstream segments of the value chain during this period. For example, costs con-verged for many polysilicon manufacturers from 2010 to 2012; one force that drove this trend was the entry of players such as South Korea's OCI Solar Power and China's GCL Solar, which contributed to polysilicon spot prices declining from about $50 per kg in 2010 to between $20 and $25 per kg today. Table 1.1 presents the largest polysilicon producers [1] in 2013.

Solar-cell and module cost curves have flattened to simi-lar degrees. As a result, value has migrated downstream to players that develop and finance solar projects and install capacity. By 2009, venture-capital firms began to shift their new solar investments from capital-intensive solar-cell manu-facturers to companies that focused on developing innovative

Table 1.1   Largest Polysilicon Producers in 2013

Company	Country of Origin	Production Capacity (Tons)	Market Share (%)
GCL-Poly Energy Holdings	China	65,000	22.0
Wacker Chemie	Germany	52,000	17.0
OCI	South Korea	42,000	14.0
Hemlock Semiconductor	United States	36,000	12.0
REC	Norway	21,500	7.0

Source: Bloomberg New Energy Finance [1].

downstream business models, such as Solar City, SunRun and Sungevity.

During the past two to three decades, the solar industry has not only marched from strength to strength but also widened its base with a significant growth such as expanding its capacity, declining its prices and improving performance attributes. Energy users have the flexibility to deploy solar-powered battery banks to dedicated loads, since no electric grid interface is required. Distributed energy generation is still largely subsidized and the solar energy technology is still a 'major investment decision' for a large section of potential investors. The growth of distributed generation depends on technological advances and its long-term viability without any subsidies at all. PV has the highest cost reduction potential and is well on its way to achieving the much desired grid parity in the current decade. Solar on-site energy generation offers a major challenge to the electric utility businesses.

1.5.2 Emerging Demand in MEA Region

SPV demand from the Middle East and Africa (MEA) region is estimated to grow 50% year-over-year in 2014. Between 2014

and 2018, annual PV demand will nearly triple as the MEA
region becomes a key market for the global industry, according to findings in the latest NPD Solarbuzz Emerging PV
Markets Report: 'Middle East and Africa'. By 2018, the annual
PV demand in the MEA region is expected to reach 4.4 GW,
with an upside potential of 10 GW. PV demand from the MEA
region in 2013 grew by 670% compared to 2012 when the
region added approximately 140 MW. Previously, the region
had a substantial share of small off-grid PV systems; however,
in 2013, the on-grid segment became the main factor driving
growth to more than 1 GW, with 1.6 GW forecast for 2014. In
2018, ground-mounted systems will account for over 70% of
the market. Until now, PV growth in the MEA region has been
predominantly driven by a small number of economically
prosperous countries, in particular South Africa and Israel.
Along with Saudi Arabia, these three countries are expected
to offer stable demand levels within the MEA region over
the next few years. In most of the Middle East countries, RE
is being seen as a means of preserving domestic oil and gas
reserves. Middle East PV demand is forecast to reach 2.2 GW
in 2018, with an upside potential of 4 GW. Israel is projected
to be the largest PV market in the Middle East in 2014, mainly
through setting up of solar parks and rooftop systems, based
on net-metering. Saudi Arabia's ambitious renewables programme will finally start to materialize this year, thus making it the largest PV market in the MEA region by 2016. Saudi
Arabia is forecast to add 2.4 GW of new PV capacity between
2016 and 2018. Other key solar PV markets in the Middle East
include the United Arab Emirates, Jordan and Kuwait.

With a severe shortage of electricity within Africa, in general, and large parts of the population of sub-Saharan Africa
having no access to electricity, solar PV is considered suitable in stimulating social and economic development in that
region. PV demand from Africa is forecast to reach 2.2 GW
by 2018, with an upside potential of 6 GW. In 2014, African

PV demand will continue to be dominated by South Africa due to the construction of large-scale PV projects resulting from the Renewable Energy Independent Power Producer Procurement Program. In the past 12 months, new plans for large PV projects have emerged across Africa, including the sub-Saharan countries of Cameroon, Swaziland and Uganda. Announcements of PV projects in the 100 MW range have now become common, as a means of quick expansion of power generation capacity. Zimbabwe is a case-specific example of this kind.

Some governments underestimate the change that is coming. SPV is a good example: total governmental projections yield less than 500 GW of SPV in 2030, but the REmap 2030 demonstrated that a combination of current market trends with enabling policies can result in 1250 GW. There is a need to focus on the overall system design rather than the often alluring consideration of locating the cheapest source of RE. Governments must ensure the development of enabling infrastructure, including power grids and storage, so as to integrate high shares of variable RE. In 2012, the United Nations General Assembly (UNGA) declared the period from 2014 to 2024 to be the 'Decade of Sustainable Energy for All', underscoring the importance of energy issues for sustainable development and for the elaboration of the post-2015 development agenda (UN GA, 2012). In the same year, the United Nations Secretary-General set up a High-Level Group on Sustainable Energy for All (SE4ALL) to develop a global action agenda based on the following three interconnected objectives:

1. Ensuring universal access to modern energy services.
2. Doubling the rate of improvement of energy efficiency.
3. Doubling the share of RE in the global energy mix (SE4ALL, 2012). IRENA is the renewable energy hub for SE4ALL.

1.6 Key Challenges and Opportunities

In the present-day scenario, the world is witnessing a major challenge of providing energy to a large section of population (estimated to be one fifth) living on earth without any access to basic electricity, thereby relying on traditional sources of fuel such as biomass. A large segment of poor population in the underdeveloped and developing world is deprived of good living conditions (e.g. lighting, cooking, education, economic development and good healthcare). Concerted efforts are being taken by governmental bodies and NGOs across the world to promote clean and efficient sources of energy especially for the upliftment of the rural people in terms of illiteracy and poverty. On the contrary, energy is available in abundance in the developed world. However, the challenges arising from fossil fuels for controlling GHG emissions and their disastrous effects on both the environment and human lives are equally insurmountable. CO_2 emissions associated with human activity and its concentration in the atmosphere have increased immensely from the pre-industrial times. The problem is compounding annually and without a major change in the structure and composition of the global energy systems.

1.6.1 Primary Global Issues at Present

- Energy prices, and high associated volatility, have become the most critical uncertainty for energy leaders for the first time this year, surpassing the global climate framework.
- The lack of global agreement on climate change mitigation remains a key issue, for the fifth consecutive year, without a clear path for the future CO_2 prices.
- Access to capital has an increased uncertainty this year, demonstrating the difficulties in matching the capital with the necessary demand for energy infrastructure.

- Carbon capture, utilization and storage is perceived with a rapidly diminishing impact, continuing the clear trend of the past 3 years and reinforcing the reality check needed around our ability to deliver on climate objectives by 2050.
- Energy efficiency remains stable in its position as an action priority for the fifth successive year and continues to present an immediate opportunity, but can only be realized with a longer-term approach to financing.

The irreversible effects of global warming such as the increase in the average temperature of the Earth's near-surface air and oceans, rising sea levels, degradation of ecosystems (that sustains life on Earth), climatic change that disturbs agriculture patterns, frequent natural calamities causing tropical diseases and energy security concerns are highly important and could not be ignored. Protecting planet Earth through a clean energy revolution calls for catalyzing low-carbon technology regimes to reduce GHS emissions for conserving nature and advancing economic growth.

The opportunities of making rapid technological advances in terms of developing highly efficient and clean energy sources are very much within the reach of scientists and technologists.

The RE sources which are available exhaustively are proving to be a viable alternative for clean power generation to fossil fuels. Drastic efforts are being taken to offset and bring new technologies to create new infrastructure and develop systems so as to suit all the energy needs. A focused approach to meet the the dual challenge of energy security and climate change is to be adopted essentially. The IPCC has reiterated that solar has the largest technical feasibility in mitigating harmful emissions from electricity production 'by a large magnitude'. For dramatically reducing greenhouse gases to the scale required to avert catastrophic climate change, the report has advocated solar energy as part of a 'portfolio of options, because no single option is sufficient'. The draft report stated that RE technologies 'hold great promise', but there are still concerns such as intermittency, (subsidized) finance and economic competitiveness, water use and land availability.

The IPCC recommendation also states that RE has the potential to more than meet global energy demand, as renewables are now the third largest contributor to global energy supply – just behind coal and gas – with a good chance of being the second largest contributor by 2020. The report also states that since 2005, solar has increased deployment by a factor of 25. Today, the cost of solar electricity generated at distributed sites like roof tops of commercial establishments, offices and residential buildings can match the cost of the electricity produced from fossil fuel, since the cost of transmission could be saved.

According to the United Nations Environment Programme (UNEP) report* by Worldwatch Institute released on 24 September 2008, efforts to curb climate change are expected to create millions of jobs worldwide in the coming decades. It has estimated investments of US $630 billion by 2030 could translate to some 20 million new jobs in RE sector, including 2.1 million in wind sector, 6.3 million in solar power sector and 12 million in biomass for energy and related industries. Further job opportunities are expected to be fuelled in large part by a corresponding growth in the market for environmental products and services worldwide, which is projected to double to US $2740 billion by 2020.

1.6.2 Markets and Policy Makers Both Play Crucial Roles

Markets provide suitable affordable solutions, but the future of sustainable energy depends on policy guidance. Policies must enable investments and stimulate market growth and

---

* UNEP Report: Green jobs towards decent work in a sustainable low carbon world produced by World Watch Institute.

transformation, with a focus not only on short-term gains but also on long-term impact. Effective policies must take into account the system and infrastructure issues, such as supply and demand vis-a-vis electricity-generation capacity. Market forces play a key role in finding efficient solutions and scaling up the best practices.

1.6.3 Five Key National Areas for Action

Five key national areas for action have been identified and are as follows:

1. Transition pathways for RE
2. Enabling businesses and knowledge
3. RE integration
4. Technology innovation
5. Enablers

1.6.3.1 Policy Measures

Targeted policies are needed to accelerate progress in these areas. There is a dire need to focus on an overall system design rather than on the cheapest source of RE. Governments must ensure the development of enabling infrastructure, including power grids and storage, so as to integrate high shares of variable RE. Pre-commercial research needs to be conducted within some emerging technology areas. Notably, new RE solutions are needed for a variety of end-use sectors. In the coming years, the international markets are set to be shaped by new technologies as well as a consequence of a certain shift in the industry's significant regions. As a result, companies dealing with solar energy will face challenges of predicting future market developments and adapting their strategies to these changing conditions as early as possible. New technologies and stiffer competition within the industry make it necessary to focus on promising markets and products. The success of future decisions thus depends on sound figures and expert assessments in no uncertain terms.

As per the report of the European Photovoltaic Industry Association, about 37 GW of photovoltaic capacity was added globally in 2013, highlighting China with 11.3 GW and Europe with 10 GW as the most important markets. The Global Market Outlook for Photovoltaics 2014–2018 reveals how markets will develop in the next few years.

1.7 Summary of Latest Indian Energy Scenario

India is the largest democracy of the world with a population of 1.24 billion according to the population census of 2011. Its GDP is equivalent to around US $18,766.80 billion and ranks 10th globally. Further it enjoys a growth rate of 5% and has a per capita gross national income equivalent to US $1571 and ranks third worldwide in terms of public–private partnerships. India possesses a liberal foreign direct investment (FDI) regime, by virtue of which it had an FDI of US $28.15 billion in 2013. The country prides itself on a vast abundance of natural resources and minerals and most importantly self-sufficiency in agriculture sector. On the energy front, India ranks as the fifth highest investor in infrastructure projects. Figure 1.3 shows a graphical representation of the GDP composition by industries. As is evident, services sector surpasses the rest of the segments in very clear terms.

Utilization of energy in India is increasing at a fast pace mainly due to its economic growth during the last decade and also by virtue of its large population. However, the country scores quite low when it comes to per capita consumption of electricity in India (2.02 kWh), which is very low compared to Canada (51.5 kWh), the United States (39.25 kWh) and other developed countries. India is one of the major coal-importing nations in the world. More than 25% of primary energy needs

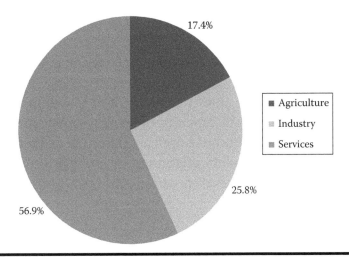

Figure 1.3   Schematics of GDP composition as represented by industry.

are being met by imports mainly in the form of crude oil and gas. India is endowed with vast RE resources including wind, solar, biomass and small hydro. India needs to develop the available RE to meet its growing power needs and ensure energy security. The total power capacity installed in the country is around 275,912 MW as on 24 August 2015. Out of this, the shares of state, central and private sectors are 34.8%, 26.7% and 38.5%, respectively. Table 1.2 highlights the key

Table 1.2   Important Achievements in the Indian Energy Scenario Capacity (in MW)

Thermal	191,664	69.5	10.83%
Hydro (renewable)	41,997	15.2	4.16%
Nuclear	5,780	2.1	5.47%
Renewables	36,471	13.2	
Total installed capacity	**275,912 MW**		

Source: Ministry of Power, Government of India website (www.powermin.nic.in) [2].

achievements of the Indian scenario in terms of energy source as on 24 August 2015.

Table 1.3 showcases the contribution of renewable energy sources to the overall Indian energy scenario as on 31 July 2015.

The off-grid/captive power capacity of SPV systems in MW_{eq} today is around 235 MW. Majority of such systems are being used to meet various end-use applications such as lighting, water pumping and battery charging (for multiple uses). India, as mentioned earlier, is rich in various RE resources, as shown in Table 1.4.

Table 1.3 Renewable Energy Technology Contribution to India's Power Supply

Wind power	23,864.91 MW
Solar power	4,101.68 MW
Small hydro power	4,130.55 MW
Biopower (biomass, gasification and bagasse cogeneration)	4,418.55 MW
Energy from waste	127.08 MW
Total installed power capacity from renewables	**36,442.77**

Source: Ministry of New and Renewable Energy, Government of India website (www.mnre.gov.in) [3].

Table 1.4 Gross Potential of Various RE Sources in MW Capacity Terms

Wind energy	49,500 MW (at 50 m hub height) 102,800 MW (at 80 m hub height)
Solar energy	50 MW/km²
Small hydro	19,700 MW
Biomass (including bagasse cogeneration)	22,500 MW

Source: Ministry of New and Renewable Energy, Government of India website (www.mnre.gov.in) [3].

1.7.1 Energy Issues and Challenges

The country fulfilled around 71% of its demand for oil in 2012 through imports, which means a very huge outgo of valuable foreign exchange. As per the available estimates, India is expected to witness a 2.5 times more power demand within the next 12 years. At present, the peak shortage of power is a modest 2%, which is against an expected energy shortage of about 5.1% in 2014–2015. But the truly alarming scenario is that around 300 million people in India still do not have access to basic electricity. Indian clean energy market is mostly driven by asset-based finance. It is the extent of 94% of the total investment in the sector. The Ministry of New and Renewable Energy (MNRE) has firmed up the 12th plan targets which are due to be achieved by March 2017 as per the figures mentioned in Table 1.5. The table also shows the respective investments for each RE energy source.

In total, it is a whopping sum of around US $33.6 billion. MNRE has in fact upgraded these targets recently so as to achieve a cumulative solar power capacity of around 1,00,000 MW by the year 2022. The national solar mission together

Table 1.5 Investment Levels Required against the Targeted RE Capacities in India

Resource	Targeted Capacity (MW)	Level of Investment Needed (in US $ Billion)
Solar power (grid connected)	10,000	12.00
Solar power (off-grid)	1,000	1.8
Wind power	15,000	2.74
Biomass-based power	2,700	2.74
Total investment required		**19.28**

Source: REINVEST 2015: A Status Report of The Ministry of New and Renewable Energy (MNRE), Government of India [4].

Table 1.6   Financial and Fiscal Incentives for RE Sector in India

Types of Incentive	Brief Remarks
Income tax holiday	• 100% for 10 consecutive years; however, minimum alternate tax • At 20% still applicable
Accelerated depreciation	• Accelerated depreciation at 80% on solar assets • Additional depreciation at 20% on new plant/machinery in the first year
Deemed export benefit	• Available to the specified goods manufactured and not actually exported • Advanced authorization from Directorate General of Foreign Trade • Deemed export drawbacks • Exemption/return of terminal excise duty
Service tax based on negative list	• Certain services are exempted from service tax • Services of transmission or distribution of electricity by an electricity utility
Customs and excise laws	• Various duty concessions and exemptions to RE sector
Reduced value-added tax	• Certain states allow reduced VAT rates (5%) on RE projects
Additional one-time allowance	• Available at 15% in the budget 2014 on new plant and machinery
Tax-free grants	• Grants received from the holding company engaged in generation, transmission and distribution of power

Source: Ministry of New and Renewable Energy, Government of India website (www.mnre.gov.in) [3].

with the solar park scheme (which is still in the draft form) are being seen as the two major RE policy initiatives in India along with several others.

1.7.2 Financial and Fiscal Incentives

It is pertinent to mention here that enhanced market outreach of the Indian RE power programme has been facilitated to a significant extent by offering several tax and regulatory incentives as shown in Table 1.6.

Simultaneously, under the non-tax incentives, the following four incentives are available in the sector of renewable energy.

■ Favourable land policies
■ Feed-in-tariffs
■ Government of India (GOI) programmes
■ Rebates

Importantly, there is 100% FDI available under the automatic route wherein no approval from GOI is required, whereas this becomes 74% in case of foreign equity participation within the purview of a joint venture. However, 100% FDI equity participation is possible with specific permission from the Foreign Investment Promotion Board (FIPB). In sum, the solar energy sector in India is expected to witness a phenomenal growth in the years to come, taking into account the new markets, expanded nature of product applications and, importantly, the fast declining cost of solar modules.

References

1. Bloomberg New Energy Finance, China: The New Silicon Valley, Polysilicon, 2 February 2015.

2. Ministry of Power, Govt. of India Website, www.powermin. nic.in.

3. Ministry of New and Renewable Energy, Govt. of India Website, www.mnre.gov.in.

4. REINVEST 2015: A Status Report of The Ministry of New and Renewable Energy (MNRE), Govt. of India.

Chapter 2

Off-Grid and On-Grid PV Applications

2.1 Introduction

Sunlight is available in plentiful measure worldwide. It is capable of lighting up homes, charging batteries, running water pumping systems and even producing power. The technology which is at work is more commonly known as solar photovoltaic or simply PV technology. Urban areas generally receive conventional grid power derived from thermal, hydro or nuclear power sources. Rural areas in contrast depend mostly on the use of non-grid power. This power is usually supplied by solar, wind, small hydro (micro and mini hydro included) and biomass sources in use individually or grouped in suitable combinations. Such combinations, for example PV–wind or PV–biomass combination, are commonly known as hybrid systems. These systems may have varying power capacities ranging between a few watts to a few kilowatts. As these systems are not connected to the grid, they are known as stand-alone-type systems or simply off-grid systems and those that are connected to the grid are known as grid-interactive/grid-connected systems. Such systems may have average power capacities ranging from a few kilowatts peak (kWp) to hundreds of megawatts peak (MWp).

The earliest demonstrated use of PV technology was witnessed around the mid-1950s in the space satellites; thereafter PV technology was universally used in low-end consumer products such as solar calculator. Since then, PV technology has moved a long way and is enabling power requirements from the milliwatts to megawatt range. Above all, the cost has been maintaining a downward pattern more so since the past 5 years. Solar power generation is now inching closer to the grid power on the cost front and is being regarded as the fastest growing power source worldwide. Currently, four successful application segments of PV technology are available:

1. *Off-grid systems* – these serve the purpose of lighting, such as in residences/communities, and are known as stand-alone systems.
2. *Off-grid industrial power systems* – these are mainly used for lighting, telecommunication and water management.
3. *Small-capacity grid-connected systems* – these systems can be set on rooftops or even integrated into building envelopes.
4. *Megawatt-capacity PV grid-power systems* – these systems are laid out in open fields for grid-connected power generation and are perceived to approach grid parity soon.

Apart from these major utilization segments, low-end consumer products such as calculators, watches, solar radios, solar caps, window ventilators, car ventilators and mobile phone chargers are continuously being sold in large numbers worldwide, particularly in countries like Japan, Korea, Taiwan and Hong Kong. Table 2.1 shows these different market segments in terms of their major end uses.

Table 2.1   Different PV Market Segments

Market Category	Range of PV Capacity	Key End-Uses
Remote villages	100 Wp–10 kWp	Lighting, battery charging, drinking water and agriculture-related needs
Remote industrial areas	1 Wp–10 kWp	Telemetry, telecom, signalization and so on
Grid-connected (small capacity)	1–100 kWp or more	Rooftops, building facades and so on
Grid-connected (large capacity)	kWp–MWp	Power utilities

2.2 Commonly Used Off-Grid PV Products

Worldwide, lighting and water pumping constitute the two most important needs of both the households and village communities. Several PV products have been developed for both indoor and outdoor applications ranging from a simple hand-held solar torch to an irrigation-based water pumping system or a refrigeration unit for a typical primary health clinic. The following section presents a bird's eye view of the most popular range of solar products finding their way mainly in rural and semi-urban areas in the first instance.

2.2.1 Solar Lantern

A solar lantern is a hand-held device which mainly comprises a solar module, battery, lamp and associated electronics. These components except the module are placed in an appropriate enclosure made of metal or plastic or even fibre glass. There are also a few lantern types with an in-built solar module. However, in majority of the lanterns, solar module is usually a detached part of the lighting unit. Interestingly, solar lanterns can be utilized for both indoor and outdoor lighting. Modern-day lanterns generally use either compact fluorescent lamp (CFL) or light-emitting diodes simply known as white LEDs.

How It Operates? Solar energy is converted into useful electrical energy by a module. Energy thus produced is stored in a sealed, maintenance-free battery for night-time use, usually ranging between 4 and 5 h.

Key Advantages:

- Around 50 L of kerosene could be saved annually by using solar lanterns.
- It is cost effective.
- It offers good-quality white light compared to faint yellow light offered by a kerosene oil lantern/wick lamp.
- It needs minimal maintenance, that is wiping clean the glass surface of the module.
- It does not emit smoke, like kerosene oil/wick lamps, and thus does not cause any kind of environmental pollution.

Key Disadvantages:

- Its use is limited.
- It is still a bit costly for those living in remote, rural areas.

Product Cost: Average cost of a solar lantern at present varies between Rs. 500 and 5000 depending on the size, make and daily hours of use.

Allowable Subsidy: The Ministry of New and Renewable Energy (MNRE) offers Central Financial Assistance (CFA) in general-category states, North-Eastern region, Andaman and Nicobar Islands, Lakshadweep Islands and special-category states. This subsidy is being made available through the National Bank for Agriculture and Rural Development (NABARD). However, it is quite important to match the minimum technical requirements as stipulated by the concerned ministry.

Market Assessment: Solar lanterns are the very first solar products made available by the Indian PV industry. A total of 9,40,000 solar lanterns have been installed in India so far. It is apt to mention here that a project on 'one million solar study lamps (comprising both CFL/LED) for empowering the populations in underserved communities' by IIT Bombay is also under progress. More lately Solar Energy Corporation of India (SECI) has gone in for the bulk procurement of solar lanterns with some multiple advantages in sight.

Potential Users: People living in residential units and apartments and people residing in remote rural villages are potential users. Patrolling parties of any nature, be it the security forces, adventure sports enthusiasts and so on, have a strong liking towards their use.

2.2.2 Solar Home-Lighting System

A solar home-lighting system is a suitable alternative for kerosene oil lamps and candle lights too. It is quite logical to move beyond the portable lighting product, that is a solar lantern. The convincing outcome is a solar home system (SHS), which is a fixed type of indoor lighting system. The simple idea is to provide lighting for a room or two in a house with a possibility of running a fan or television too.

How It Operates A solar home-lighting system is run by freely available solar energy via solar modules. Electricity thus generated is stored in a battery as in the case of a solar lantern. The only difference is that the battery is of a higher capacity and not an in-built component. Importantly, charge controller is also included as it prevents overcharging and deep discharge of the battery.

Key Advantages:

- It is a very good option for areas without any regular grid power supply.

- It is possible to save around 100 L of kerosene oil per system (by using a 60 Wp solar module) on an annual basis.
- It is possible to avail low-cost financing from banks.

Key Disadvantages:

- Battery storage is expensive if a higher capacity system is to be designed.
- Initial capital cost of the system is still high for those who belong to low economic status.

System Cost: Solar home-lighting system with a module power capacity of 10–100 W costs between Rs. 6,000 and 45,000. This includes the cost of installation and a year of product service.

Applicable Subsidy: MNRE offers CFA in general-category states, North-Eastern region, Andaman and Nicobar Islands, Lakshadweep Islands and special-category states. This subsidy is being made available through the NABARD. Importantly, the system has to match the technical requirements as specified. SHSs are expected to have a payback period of nearly 2 years at present.

Market Assessment: SHSs are best suited to locations that are devoid of any grid-power supply. A total of 10,01,890 systems have been deployed across India till date. Solar systems of this type are also effective in urban regions as they are dependent on inverters for backup power supply.

Potential Users: Individual residences, small business establishments, educational institutions and panchayats are some of the potential users.

2.2.3 Solar Street-Lighting System

Solar PV (SPV) systems are capable of meeting both the indoor and outdoor lighting requirements. Even today, it is not uncommon to see urban streets in pitch dark conditions, not to mention about the remote rural areas. Solar street-lighting systems are stand-alone systems driven by the abundantly available solar energy.

How It Operates? Solar energy is converted into useful electricity and stored in a battery. A unique feature of solar street-lighting system is that it switches ON and OFF automatically between dusk and dawn every day. In this case, the solar module itself acts as a light-sensing device. The solar street-lighting system is a pole-mounted system and spreads the light uniformly.

Key Advantages:

■ Solar-activated sensor can also be used to switch ON and OFF the conventional power-based lighting systems. Energy can thus be saved.
■ Solar street-lighting systems are quite beneficial both in the urban and semi-urban settings.
■ It makes rural surroundings safer, which are vulnerable to intrusion by animals at times, especially those areas that are in close proximity of the forests.

Key Disadvantages:

■ Initial capital cost of the solar street-lighting system is still high.

System Cost: The power capacity of a solar street-lighting system ranges between 35 and 150 Wp. The approximate cost of a 75 Wp solar street system which includes the cost of a solar module ranges between Rs. 10,000 and 15,000 at present.

Applicable Subsidy: MNRE provides CFA in general-category states, North-Eastern region, Andaman and Nicobar Islands, Lakshadweep Islands and special-category states. This subsidy is being made available through NABARD. Importantly, technical requirements are to be matched with those stipulated by MNRE.

Market Assessment: Solar street-lighting systems are extremely useful in urban and semi-urban settings due to long power cuts at night. Usage of street lights has increased the safety of villages, especially those surrounded by forests. Importantly, it has also led to an extended hours of community participation beyond the dusk. Likewise, industrial units located in power-deficiency areas can turn to the use of these very low-maintenance stand-alone systems in far larger numbers. A total of 2,74,679 solar street lights have been installed in India so far. Amongst the most potential end-use segments are the residential societies, municipalities, institutions, industries, panchayats and so on.

Potential Users: These light systems are primarily used in rural, semi-urban and urban settings for unhindered illumination during night-time and enhanced security.

2.2.4 Solar Fans

India is blessed with plenty of sunlight and thus is a high-temperate zone. Remote rural area populations have to be content with a small fan due to the paucity of higher amount of solar power. A solar fan is a mechanical device run by a small-capacity solar module.

How It Operates? Solar fans are normally direct current (DC) fans unlike conventional alternating current (AC) fans used in grid-connected areas. It is also possible to make use of an AC fan, but a solara inverter is required for its operation. Solar fans are directly used during daytime and can be run with a battery storage at night.

Key Advantage: The electrical motor of a solar-powered fan hardly produces any noticeable noise.

Key Disadvantage: Solar DC fans of higher capacity would involve the use of much higher capacity solar modules and thereby incur higher initial capital cost.

Product Cost: Average cost of a solar fan varies between Rs. 3000 and 5000. This includes the cost of solar modules, storage battery and so on.

Allowable Subsidy: MNRE does not provide any kind of subsidy exclusively for encouraging the use of solar fans. However, as solar home-lighting systems can also run a solar fan, subsidy can be obtained in an indirect way by availing fund for solar home-lighting systems.

Market Assessment: Rural markets especially with little or no supply of electricity may opt for this low-power rating fan.

Potential Users: Users of solar home-lighting systems, especially within a non-grid-connected area, will also need solar fans.

2.2.5 Solar Garden Lights

Solar lighting products come in varieties, shapes, sizes and makes. Solar garden light is a good example, which is an aesthetically pleasing device and can be placed easily anywhere. It is made of a mini solar module, an LED lamp and a small built-in battery. Solar garden lights are available in different types such as spot lights, hanging lights, ground stake lights, landscape lights and lamp post lights. A unique feature of this type of lights is the presence of motion sensors, which can be used for security purposes.

How It Operates? Sunlight charges a built-in battery and the automatic light sensor switches on the garden light after the sun sets. The LED lamp offers good-quality light for several hours at a stretch.

Key Advantages:

- ▪ It can be easily installed at any point of use.
- ▪ It occupies the smallest possible area and thus can be put up in a good enough number anywhere.
- ▪ Its full range of components easily fit in a built-in assembly.

Key Disadvantages:

- It is still a bit expensive than the conventional type of garden light. Average price ranges between Rs. 300 and 1200 based on the encasing body material, such as steel, plastic or carbon fibre.

Product Cost: The price of an individual unit varies between Rs. 300 and 1200.

Market Assessment: Solar garden lights are well suited for apartments, gardens, parks, farmhouses, stadiums, educational campuses and so on.

Potential Users: These types of solar garden lights can be made use of in public parks. In addition, due to their novelty and aesthetic considerations, these lights may also be used in hotels, tourist cottages, farm houses and so on.

2.2.6 Solar Chargers

A number of both electrical and electronic devices currently run on rechargeable type of batteries. The most prominent of these is the mobile telephone. Such a battery can normally be charged using an AC socket. However, while in transit, a solar charger can meet the same need easily. A solar charger can well be regarded as a handy alternative to the existing range of electrical chargers.

How It Operates? Sunlight available on a daily basis is absorbed by a mini solar module. The power thus made available is used to charge the low-capacity battery in a mobile phone, for example. However, larger capacity is needed for charging a high-capacity battery such as in a laptop.

Key Advantages:

- Being lightweight, it can be easily taken to any location for charging purposes.
- It is very cost effective.

Key Disadvantages:

- It cannot be used in the absence of sunshine.
- It is still expensive when compared to a conventional electric charger.

Product Cost: Average cost of a solar-run mobile phone charger is between Rs. 500 and 4500. However, a solar laptop charger may cost between Rs. 7,000 and 10,000, depending on the capacity of the solar module used.

Allowable Subsidy: There is no direct subsidy available to the users.

Market Assessment: There is a huge market potential for such appliances as an extremely large number of people use mobile phones. In fact anyone who has a regular need for charging and does not wish to go in for frequent purchase of expensive batteries can take an easy recourse to the solar battery charging.

Potential Users: It is handy and beneficial to all who keep travelling. It is also of great value in situations when the conventionally charged battery backup does not last long due to heavy usage.

2.2.7 *Solar Inverter*

Solar power is a DC source of power. This clearly implies a need to convert it into usable AC electricity so as to run the normally available AC appliances at homes. The alternating current thus produced is then fed to a commercial grid. From the utility grid, AC power is supplied to residential buildings. The device that helps in this conversion is commonly known as a 'solar inverter'. Key components of a solar off-grid inverter include solar modules, charge controller, batteries and inverter.

How It Operates? The following are a few principal modes by which solar inverters operate.

2.2.7.1 Off-Grid Mode

Simply put, a home inverter utilizes the DC power available from the batteries and converts it into AC power for usage. The function of a solar inverter is more or less the same if it is an off-grid solar power system. The system is designed in such a way that it depends either fully on solar power or on the batteries that are charged from solar power when sunshine is available. The same system can switch to the grid system during the no-sunshine period of the solar modules.

2.2.7.2 On-Grid Mode

A grid-connected solar power system is mainly expected to supply surplus electricity to the grid. The system basically converts the DC power into more usable AC power. However, it brings in an additional requirement of an MPPT (maximum power point tracking) or PWM (pulse width modulation) feature to be built in it. This is because the voltage produced by a solar module may change much in accordance with the available temperature and solar insolation. Thus, MPPT feature enables the extraction of the maximum power from a solar module. There is also an additional feature available in grid-connected solar inverters, that is, anti-islanding. The purpose is to prevent the presence of any solar power at the time the grid fails. This is helpful in safeguarding technicians who may be involved in repairing any line at the moment the grid fails when the solar system would still be producing the power.

Key Advantages:

- Higher efficiency and enhanced field performance reliability.
- Simple installation and easy maintenance.

Key Disadvantages:

■ These still have a high initial capital cost.

Product Cost: The cost of a solar inverter is heavily dependent on its capacity. Added to it is the cost for the purchase of solar modules and batteries.

Allowable Subsidy: The Ministry of New and Renewable Energy has designated NABARD to make available subsidy for the purchase of solar inverters. It is being enabled under the Jawaharlal Nehru National Solar Mission (JNNSM) launched by the Government of India in 2010.

Market Assessment: Solar inverters are potentially used in residential buildings, commercial buildings and institutions.

Potential Users: Solar inverters are of potential use as one cannot solely depend on the AC mains for charging the battery. There are several areas within which the grid power interruption is quite high at times, thus forcing the users to adopt viable alternatives like a solar inverter.

2.2.8 SPV Water Pumping System

Lighting, battery charging and pumping are widely recognized as being the most preferred uses of solar PV technology worldwide. In India, small-capacity water pumping systems mainly for drinking water requirements were introduced in the late 1980s. A solar water pumping system is defined as an electric pump driven by solar power. It mainly comprises a PV array, motor pump set unit, mounting structure and interconnecting wires and cables. PV array is mounted on a stand and connected to a motor pump set. As the sunrays fall on the solar modules, DC electricity is generated and is used to run a DC motor pump set or an AC motor pump set. An AC motor pump set requires an inverter.

How It Operates? Solar modules forming a PV array are generally connected in a series and parallel arrangement to produce DC power. The motor is energized by such power and drives the pump. This pump then makes the water flow under some pressure.

2.2.8.1 Key Solar Water Pumping Technologies

Water pumping is not a new application. However, what is relatively newer is the use of freely available sunlight to operate the motor pump unit. Centrifugal pumps and positive displacement pumps are the two main types of solar water pumping technologies. In centrifugal pumps, the pumps rotate at high speed and water is sucked in through the middle of the pump. Majority of AC pumps use such a centrifugal impeller. In contrast, a good number of DC-powered pumps use centrifugal force or positive displacement principle so as to move the fluid. At present, this type of pump is mostly being used. The pump moves water into a chamber and then forces it out via a piston on a helical screw.

Types of Commercially Available Solar Pumps:
The following few are the most important types of solar water pumps available in the country.

  Surface pumps – This is normally positioned beside water sources such as lake, water, stream and canal.
  Submersible pumps – This is submerged in water sources such as a deep well.
  Floating pumps – This is usually placed on the top of the water source.

Surface pumps are cheaper than the submersible pumps. However, their suction capacity is limited with a suction head of around 6–7 m.

Key Advantages:

- Solar pumps easily replace the use of imported fuel commodity, for example the diesel.
- They do not pollute the environment.
- Operative cost is less compared to diesel pumps.
- Solar pumps when not in use during the lean season can be put to other uses. This simply means that SPV arrays are also used in household needs such as lighting.

Key Disadvantages:

- Solar pumps still have a high initial capital cost.
- The water output varies as per the solar radiation intensity.

Product/System Cost: Solar water pumping systems are currently available in different capacities and makes. The cost of 1 HP pumping system generally ranges between Rs. 2 and 2.5 lacs as against the costs mentioned here for other HP categories.

Pump Capacity (HP)	Cost (Rs. lacs)
2	3.0–3.5
3	4.25–4.75
5	7.25–7.75

Allowable Subsidy: CFA is available from MNRE in respect of the states/regions mentioned above. Further this subsidy is available through NABARD as per Table 2.2.

Importantly, solar water pumping systems have to meet the technical requirements as stipulated by MNRE.

Cost Analysis: It costs between Rs. 17 and 20 for every unit of electricity produced from a diesel generator for water pumping

Table 2.2   Cost Estimate of Various PV Pump Configurations

Pump Type	Pump Capacity	Cost
DC pumps	Up to 2 HP	Rs. 57,600
DC pumps	More than 2–5 HP	Rs. 54,000
AC pump	Up to 2 HP	Rs. 50,400
AC pump	More than 2–5 HP	Rs. 43,200
AC pump	More than 5–10 HP	Rs. 38,800

Source: Ministry of New and Renewable Energy, Government of India website (www.mnre.gov.in).

application. Solar water pumping is relatively cheaper at Rs. 10–13 per unit. If a solar pump is delivered free of any subsidy, the payback period generally ranges between 6 and 8 years. However, with subsidy, this period is reduced to 4–6 years.

Market Assessment: The Government of India has recently allocated a large sum of around Rs. 400 crores for the installation of about 10,00,000 solar water pumps for a diverse range of applications.

- To supply drinking water in villages, schools, hospitals, homes, animal farms and poultry besides residential societies.
- To supply water for irrigation purposes in fields, farms, greenhouses besides drip- and sprinkler-based irrigation for the agro-based industries.

The Indian PV water pumping for agriculture and related needs was launched in 1993 with a targeted deployment of around 50,000 pumps. However, only about 11,626 solar pumps could be installed so far.

2.3 Grid-Connected Rooftop Solar Power Plant

Recently, Jawaharlal Nehru National Solar Mission (JNNSM) has catalyzed the growth of solar market in India. The Government of India is now committed towards achieving 100 GW of grid-interactive solar power capacity by 2022, out of which, 40 GW would be deployed by decentralized and rooftop scale solar projects alone. Rooftop SPV is expected to play a significant role in meeting energy demands across the diverse segments of energy use. Such systems, especially for use in industrial and commercial sectors, are already turning out to be cost competitive. The estimated realistic market potential for rooftop PV in urban settlements of India is about 124 GW according to a mapping study undertaken by the Energy and Resources Institute.

MNRE announced the rooftop PV and small-scale solar power generation programmes under phase I of the solar mission so as to encourage grid-connected projects. Recently, the quantum of allowable subsidy has been brought down from 30% to 15%. The use of rooftop PV systems in a capacity range 100–500 kWp is also encouraged by the SECI under the RESCO model. Under this model, SECI has offered to purchase and sell electricity to the utilities at Rs. 5.50/kWh. The Central Electricity Authority (CEA) is the nodal agency responsible for the technical guidelines regarding the grid interconnection of small-capacity-distributed generating systems including the rooftop systems.

2.3.1 Role of State Governments

MNRE is rightly supported in the smooth implementation of its various programmes by the state nodal agencies for renewable energy. Various states have also launched their state-specific policies for rooftop SPV programmes. In fact, both the off-grid and on-grid PV rooftop projects are being implemented by state governments. Importantly, State Electricity Regulatory Commissions (SERC) that comprises 17 different states has already issued relevant regulatory orders for grid-connected rooftop PV projects. Gujarat is the first state in India to implement a rooftop PV pilot scheme. This scheme permits direct sale of solar-generated electricity to the utility grid.

2.3.1.1 Institutional Arrangements

A wide range of stakeholders are involved in the wholesome development of PV programmes in India. MNRE is the nodal organization responsible for policy formulation. Solar Energy Corporation of India and NVVN are some of the central nodal agencies involved in this programme. State nodal agencies deal with issues at the respective state government levels. As for the regulatory framework, the central electricity commission and state electricity regulatory commissions meet this key responsibility. The responsibility of resolving technical issues rest solely with the CEA and issues related to net metering are resolved by DISCOM. As solar power is still costly, funds are made available through the Indian Renewable Energy Development Agency (IREDA), public sector banks, private banks and, importantly, at times also by bilateral and multilateral organizations. Solar developers, equipment suppliers and system installation agencies also play a significant role in this programme.

2.3.2 Brief Outlook on Global PV Rooftop Market

SPV rooftop programme is making faster market inroads in many countries of the world. Notable amongst these are the programmes being run in Germany, the United States, Japan, Italy, Australia and India. Several types of policy and financial and fiscal incentives are in place so as to boost this specific PV segment which is envisaged to be full of commercial promise. Table 2.3 shows the PV-specific incentives available in selective countries.

2.3.2.1 Rooftop System Description

SPV rooftop is a roof-mounted system where solar panels are mounted on the roof of a building complex. The major components of a SPV rooftop system are as follows:

■ PV modules
■ Battery bank

Table 2.3 PV-Specific Incentives Available in Selected Countries

Scheme	Australia	Germany	India	Italy	Japan	USA
Direct capital subsidy		√	√	√	√	√
Green electricity schemes	√	√		√		√
PV-specific green electricity scheme	√	√	√	√		√
Renewable portfolio standard (RPS)	√	√	√			√
Solar set aside RPS target			√			√
Financing scheme	√	√	√			√
Tax credits/tax benefits	√	√	√	√	√	√
Net metering/net billing/self-consumption incentives	√	√	√	√	√	√
Sustainable building requirements					√	√

Source: Solar PV Rooftop Systems in India 2015, Shakti Foundation Report, www.shaktifoundation.in/reports [1].

- Mounting structure
- Inverter
- Junction box cables and other electrical accessories
- Charge controller

A typical SPV rooftop installation can be of any one of the following types: grid-connected, grid-connected with storage or off-grid.

2.3.2.1.1 Technology Description

When the charged particles of sunlight called photons fall on the solar cells in the solar panel, they energize the electrons in the solar cell, thereby producing electricity. The electricity so produced is in the form of DC. The DC supply charges a battery, which is connected with the solar rooftop system. The DC supply can either be used directly to run DC appliances or can be converted into AC supply by using a solar inverter or a solar power–conditioning unit (PCU). The AC supply produced from the solar rooftop system can be used to run household appliances.

- PV modules must conform to the latest edition of any of the following IEC equivalent/BIS Standards for PV module design qualification and type approval: Crystalline Silicon Terrestrial PV Modules IEC 61215/IS 14286.
- Thin Film Terrestrial PV Modules IEC 61646/Equivalent IS (under development).
- Concentrator PV Modules and Assemblies IEC 62108.
- In addition, the modules must conform to IEC 61730 Part 1 requirements for construction and Part 2 requirements for testing, for safety qualification or Equivalent IS (under development).
- PV modules to be used in a highly corrosive atmosphere (e.g. coastal areas) must qualify Salt Mist Corrosion Testing as per IEC 61701/IS 61701.

2.3.2.2 Identification and Traceability

Each PV module must use a radio-frequency identification tag, which must contain the following information:

- Name of the manufacturer of PV module
- Name of the manufacturer of solar cells
- Month and year of the manufacture (separately for solar cells and module)

- Country of origin (separately for solar cells and module)
- Current–voltage characteristics, that is, I–V curve for the module
- Peak wattage, Im, Vm and FF for the module

Beneficiaries: Individuals, industrial/commercial/non-commercial entities are benefitted.

Average Size: Typical size of a solar rooftop system may generally vary between 1 and 500 kWp. However, currently such rooftops are also available which accommodate megawatt scale power plants, especially for industrial and institutional purposes.

Authorized Testing Laboratories/Centres: PV modules must qualify (enclose test reports/certificate from IEC/NABL accredited laboratory) as per relevant IEC standard. Additionally, the performance of PV modules at standard test conditions (STC) must be tested and approved by one of the IEC-/NABL-accredited testing laboratories including Solar Energy Centre of MNRE. For small-capacity PV modules, up to 50 Wp capacity STC performance as mentioned earlier, will be sufficient. However, qualification certificate from IEC-/NABL-accredited laboratories as per relevant standard for any of the higher wattage regular module should be accompanied by an STC report/certificate.

   Solar rooftops of the modern era are different from the stand-alone rooftop systems of the past. In a grid-connected area, rooftops offer a paradigm shift in terms of use of what is now commonly known as the net metering. The following section takes a close look at the net metering and its associated aspects in brief.

2.4 Solar Net-Metering

In case of grid-connected SPV systems, solar energy produced by the solar panels is converted to AC supply by a solar grid inverter. The output of the solar grid inverter is connected to

Distribution network

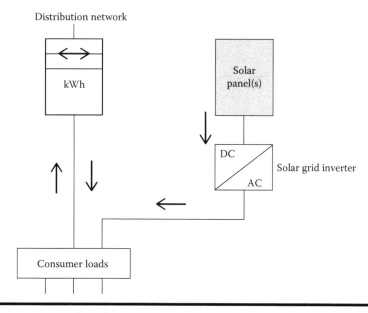

Figure 2.1 Simple schematic layout of a solar PV rooftop system.

the distribution switch board (Figure 2.1) of the building. The electrical energy flows to the loads of the buildings (e.g. lights, fans and appliances). Excess solar energy produced will automatically be exported to the distribution network (the grid). If the solar energy produced is less than what the loads of the building require, then the shortfall will be drawn from the grid (energy import).

The Service Connection Metre: For installing a solar net metre the existing service connection should be replaced with a metre that can measure both energy import (from the grid to the consumer) and energy export (from the consumer to the grid). These meters are commonly known as bidirectional energy meters or import–export energy meters.

Space Requirement: The size of area required for SPV rooftop varies depending upon the capacity of the installation. The area must be free of any shade so as to allow the maximum solar insolation on the rooftop system. A typical capacity of 1 kWp (kilowatt peak) system, if placed in a single row, would require an approximate shade-free area of 80 ft² (around 8 m²). The space required for 1 kWp system would be about 120 ft² (around 12 m²), if panels are placed in multiple rows. This increase in space requirement is to avoid shadow of one row from falling over the other. It is suggested that the solar panel be installed facing south so that it absorbs the maximum solar radiation and is more efficient. SPV rooftop systems are also available with a tracking mechanism, wherein a solar tracker device orients the PV panel(s) according to the movement of the sun throughout the day. This significantly increases the absorption of solar radiation power output from the PV panel(s).

Key Disadvantages:

- Intermittent nature of solar energy generation due to diurnal and seasonal variations.
- Off grid system requires a battery to store the additional electricity that is generated by the rooftop SPV system. This leads to an additional set-up cost.
- The area required by the PV panels is large.

System Cost: The estimated cost of a typical 1 kWp off-grid (with battery backup) SPV rooftop system is about Rs. 1 lac whereas that of a grid-connected system is Rs. 0.75 lacs. The price depends on technology, user requirements and so on.

2.4.1 Cost Analysis

More than 50% of the cost of rooftop PV system is spent on the solar panels and 20% on the inverters. A 15% is spent on components such as wires, charge controllers and mounting auxiliaries, which are called the balance of system components. The remaining percentage of the total cost is spent on the installation of the entire system. Battery efficiency and electricity backup period add up heavily to the cost of the system. Batteries add to the initial cost, recurring maintenance and replacement expenditure. A battery backup would add up to 12.5% to the total cost of the system.

2.4.2 Key Objectives of the PV Rooftop Programme

- To promote the grid-connected SPV rooftop and small SPV power-generating plants amongst the residential, community, institutional, industrial and commercial establishments
- To mitigate the dependence on fossil fuel–based electricity generation and encourage environment-friendly solar electricity generation
- To create an enabling environment for investment in solar energy sector by private sector, state government and the individuals
- To create an enabling environment for supply of solar power from rooftop and small plants to the grid
- To encourage innovation in addressing market needs and promoting sustainable business models and ensure employment opportunities
- To provide support to channel partners and potential beneficiaries, within the framework of boundary conditions and in a flexible demand-driven mode
- To create a paradigm shift needed for commoditization of grid-connected SPV rooftop applications
- To support consultancy services, seminars, symposia, capacity building, awareness campaigns, human resource development and so on
- To encourage replacement of diesel, wherever possible

2.5 Electricity Bill with Solar Net-Metering – A Few Case-Specific Examples

With solar net-metering, the consumer pays for the difference between the import and export energy (the net-metered energy). For example, during a billing cycle, a consumer imports 900 kWh (units) and exports 500 kWh. Then the electricity bill will be only for 400 kWh. If the export energy exceeds the import energy, the excess of import kWh will be carried over to the next billing cycle. During a 12-month period (the settlement period), the maximum of energy export that will be credited by the distribution company is 90% of the energy import. For example, during the settlement period, there is a total import of 5000 kWh and a total export of 6000 kWh. Of the 6000 kWh exported, 4500 kWh is eligible for adjustment with the import kWh (90% of 5000 kWh).

SPV System Capacity: The SPV system capacity should not be more than the approved load of the service connection. It is also advisable to have an SPV system size that has an annual estimated generation of not more than 90% of the estimated consumption (see above).

2.5.1 Getting a Solar Net-Metering Connection – Five Convenient Steps

Step 1 – Application
Make an application to the Office of the Executive Engineer (O&M) of the distribution company of your area. The executive engineer of the operation and maintenance unit (O&M) acts as a nodal officer for solar net-metering. Your application will be registered in a computerized database and you have to pay a registration fee of Rs. 100.00. You will get a signed acknowledgement. For high-tension (HT) service connections, the application should be registered by the superintending engineer of the distribution circle.

Step 2 – Technical Feasibility
The distribution company will verify the technical feasibility of connecting your SPV system to their distribution network. This is done based on the following two criteria:

- The total SPV capacity in the local distribution network (existing and proposed) should not exceed 30% of the distribution transformer capacity. For example, a distribution network that is served by a 250 kVA distribution transformer cannot have more than 75 kW of total SPV capacity connected to it.

■ The proposed SPV system capacity should not be more than the approved or contacted load of the service connection. If the proposed SPV capacity is more than the approved load of the service connection, then you have to first apply for the load enhancement of the service connection and then for your solar net-metre connection.

If the distribution company finds that your proposed SPV system can be connected to their local distribution network, you will receive from them a technical feasibility intimation letter. This letter will be sent to you within 10 working days from the date of your application.

Step 3 – SPV System Installation and Readiness Intimation
You can now procure and install the SPV system. This has to be done within 6 months from the date of the technical feasibility intimation letter. This period can be extended to another 3 months upon written request if system procurement has been completed and installation work is in progress. Upon possible completion of the SPV system installation work, you intimate your readiness to the executive engineer (O&M) of the concerned distribution company.

Step 4 – Safety Inspection
Within 10 days from the date of receiving your readiness communication, your SPV system will be inspected for safety. For SPV systems up to 10 kW, this will be done by the distribution company, and for systems above 10 kW, this will be done by the electrical inspector of your area. Within 5 days from the date of inspection, you will receive a safety certificate if the installation complies with the technical requirements.

Step 5 – Service Connection Metre Replacement and Commissioning
The distribution company will replace the existing service connection metre with a bidirectional metre for which you have to pay metre replacement charges. TANGEDCO permits the consumer to procure bidirectional metre, which has published on its website a list of approved metre makes and types.

2.5.2 Safety Requirements for Grid-Connected SPV Systems

1. The solar grid inverter stops feeding power into the loads and the distribution company grid when the grid fails or is switched off for maintenance. For this purpose, the solar grid inverter has built-in anti-islanding protection. This protection ensures that the solar grid inverter does not operate in island mode. The anti-islanding protection can be tested by switching off the main connection, or by removing the metre fuses; then check whether there is voltage on the consumer side of the main connection or metre fuses. If there is no voltage, then the anti-islanding protection is functional.
2. The SPV system has its own separate earthing system and is provided with lightning protection.
3. It is advisable (not mandatory) to install surge protection devices on both the DC side and the AC side of the solar grid inverter.
4. A caution sticker of 10 × 7 cm should be fixed by the consumer on the main switch and near the metre with the text 'Solar PV System'. The letters should be white on a green background. The distribution company will fix similar stickers on the service connection pole and elsewhere in the distribution network. Additionally, a caution sticker should also be fixed near the service connection metre with the following text: 'This service is fitted with an LT grid connection'.

In a grid-connected rooftop or small SPV system, the DC power generated from SPV panel is converted to AC power with a PCU. It is fed to a grid either with 33/11 kV three-phase lines or with 440/220 V three-/single-phase line depending on the capacity of the system installed at the institution/commercial establishment or residential apartments and the regulatory framework specified for respective states. They generate power during the day time which is utilized fully by powering captive loads and feed excess power to the grid as long as the grid is available. In case solar power is not sufficient due to cloud cover, the captive loads are powered by drawing power from the grid. The grid-interactive rooftop system works on a net-metering basis, where the beneficiary pays to the utility based on the reading of net metre.

Alternatively 2 m can also be installed to measure the export and import of power separately. Mechanisms based on gross metering at a mutually agreed tariff can also be adopted. Rooftop grid power plants can be installed on the roofs of residential and commercial apartments, housing societies, community centres, government organizations, private institutions and so on. Ideally, grid interactive systems do not require a battery backup as the grid itself acts as the backup for feeding excess solar power and vice versa. However, to enhance the performance reliability of the overall systems, a minimum battery backup of 1 h of load capacity is recommended. In grid interactive systems, it must however be ensured that in case the grid fails, the solar power has to be fully utilized or it should be stopped immediately from being fed to the grid (if any in excess); this is, for example, to safeguard any grid person/technician from getting an electric shock (electrocuted) while working on the grid for maintenance. This feature is termed as 'anti islanding protection'. The grid-connected rooftop SPV generation plants generate electricity at the consumption centre and hence reduce the network losses of the distribution licensee. The electricity generation also contributes to meet the demand and supply gap and also enables the obligated entities for complying with their solar purchase obligation targets as specified by appropriate Electricity Regulatory Commissions. India has a huge potential for deployment of grid-connected rooftop SPV power generation plants and the MNRE envisages harnessing this potential.

2.5.3 Technical Standards for Connectivity

CEA is an apex organization which functions under the administrative control of the Indian Ministry of Power. The following set of technical guidelines [2] and protocols related to grid connection of PV systems has been prepared by CEA.

1. *Connectivity Regulations*: CEA has notified 'CEA (Technical Standards for Connectivity of the Distributed Generation Resources) Regulations, 2013' [2]. This standard provides necessary guidance to distribution licensee/ DISCOMs and also the transparency in the process and encourages consumers for installing grid-connected rooftop solar plants.
2. *Tariff Determination*: The projects can be installed based on net metering or feed-in-tariff (FIT). This will be decided by Regulators/DISCOMs/Distributed Licensee in consultation with the implementing agencies. In case of FIT, the provisions should be in such a manner that a safeguard is provided to all stakeholders including DISCOMs. The tariff is attractive for the roof owner and does not put too much burden on the DISCOMs. Therefore, regulators have to come up with FIT for rooftops with and without MNRE subsidy.
3. *Availability of Electricity Grid*: An electricity grid should be available near the solar installation facility; if not, it has to be provided by the concerned agencies.

4. *Signing of MoU/Agreements*: If required, a Memorandum of Understanding (MoU) has to be signed among the beneficiaries/DISCOMs/distribution licensees and the other involved parties.

2.5.3.1 Classification of Projects Based on Grid Connectivity

The projects under these guidelines fall within two broad categories: (1) the projects connected to HT voltage at distribution network (i.e. below 33 kV) and (2) the projects connected to LT voltage, that is 400/415/440 V (three phase) as the case may be or 230 V (single phase). Accordingly, the projects may be divided into two categories.

Category 1: Projects connected at HT level (below 33 kV) of distribution network
The projects with proposed installed capacity of range 50–500 kW and connected at below 33 kV fall within this category. The projects should follow appropriate technical connectivity standards in this regard.

Category 2: Projects connected at LT level (400 V – three phase, or 230 V – single phase)
The projects with proposed installed capacity of less than 100 kW and connected to the grid at LT level (400/415/440 V for three phase or 230 V for single phase) fall within this category.

2.5.3.1.1 Procedural Mechanism/Submission of Proposals

The project site/rooftops at office buildings, commercial buildings, residential apartments and so on are selected on the basis of total energy requirement of the premise and the area available for installation of rooftop SPV system. SPV systems installed on the rooftop of selected buildings should meet the requirement of the building as much as possible in agreement with the local DISCOMs/distribution licensee. Though rooftop systems shall be generally connected to low voltage (LV) supply, large SPV system may be connected to 11 kV system. The voltage level in the distribution system for ready reference of the solar suppliers is selected based on the following criteria; however, the connectivity level may be decided depending upon site conditions and policies.

■ For SPV systems with a capacity of up to 10 kW, the connectivity is by low-voltage single-phase supply point.
■ For those with a capacity between 10 and 100 kW, the connectivity is by a three-phase low-voltage supply point.
■ For systems above 100 and up to 500 kWp capacity, connection can be made at 11/33 kV level.

Export–import metres/two-way/bidirectional metres can be installed with the facility of net metering. Two-way metres can also be used as they are cheaper and give a better idea about the power exported. The metre may also be finalized in consultation with the distribution licensee/DISCOM. The CEA regulations on metering arrangements are as follows:

■ The billing of buildings by DISCOM can be done on the basis of net energy drawn from the grid during the month on the tariff prescribed by the Regulatory Commission for commercial consumers or as finalized with the DISCOM.
■ A Power Purchase Agreement has to be signed between the owner of the building, third party and the DISCOMs as applicable. If the state has already announced a policy on the grid-connected rooftop and small solar plants, the relevant notification may be mentioned along with the proposals.
■ An agreement between DISCOM and the owner of building/premise/SPV plant has to be signed for the net metering and billing on a monthly/bimonthly basis as applicable. Suitable payment security mechanism has to be provided by the DISCOM/distribution licensee/state nodal agency/utility and so on.

2.5.4 Business Models for Grid-Connected Rooftop and Small Solar Power Plants

For a smooth operation of rooftop and small solar power plants, various situations and conditions based models may be worked out so as to make it a workable business model. The business models must be in accordance with the prevailing legal framework. Many business models are possible and some of them are as follows:

1. *Solar installations owned by consumer*
   a. Solar rooftop facilities owned, operated and maintained by consumer(s). Solar rooftop facilities owned by consumer but operated and maintained by the third party.
2. *Solar installations owned, operated and maintained by third party*
   If a third party implements the solar facility and provides services to the consumers, surplus electricity may be fed to the electricity grid. The combinations could be as follows:
   a. Arrangement as a captive generating plant for the roof owners
   The third party implements the facility at the roof or within the premise of the consumers; the consumer may or may not invest as equity in the facility as mutually agreed between them. The third party may also make arrangements for undertaking operation and maintaining the facility. The power is then sold to the roof owner.
   b. Solar Lease Model, Sale to Grid
   The third party implementing the solar facility shall enter into a lease agreement with the consumer for a medium- to long-term basis. The facility is entirely owned by the third party and consumer is not required to make any investment. The power generated is fed into the grid and the rooftop owner gets a rent.

3. *Solar installations owned by the utility*
   a. Solar installations owned, operated and maintained by the DISCOM

   The DISCOM may own, operate and maintain the solar facility and also may opt to sub-contract the operation and maintenance activity. The DISCOM may recover the cost in the form of suitable tariff. The electricity generated may also be utilized by DISCOM for fulfilling the solar renewable purchase obligation.
   b. Distribution licensee provides appropriate viability gap funds

   The DISCOM may appoint a third party to implement the solar facilities on its behalf and provide appropriate funds or viability gap funds for implementing such facility. It may also enter into an agreement with the third party undertaking the operation and maintenance of the solar facilities. There can be many such business models which may be developed/adopted depending upon the market conditions, user's interest and initiatives by the ESCOs.

2.6 Estimated Rooftop Potential for Central Government Buildings

A country-wide exercise has been undertaken to identify government buildings that can house a rooftop PV system. Underlying assumptions include the following few:

- Around 40% of the identified roof space can be used for rooftop solar systems.
- Roof has enough structural load-bearing capacity to support the solar system.
- Sufficient capacity is available at the distribution transformer level.

Table 2.4 presents the estimated rooftop potential of around 7196.40 MW within the buildings housing the Ministries/Departments of the Government of India.

2.6.1 Current Status of PV Rooftop Deployment in India

As of now, solar rooftop projects with a cumulative capacity equivalent to 360.81 MW have been sanctioned by MNRE. Of this, a capacity of 54.187 MW stands commissioned and is shown in Table 2.5.

Table 2.6 presents the state-wise break-up of the Status of Grid Connected SPV Rooftop Projects Sanctioned to States/Uts/SECI/PSUs and Other Government Agencies.

Till date, 13 states have come out with solar policy supporting grid-connected rooftop systems. Further state electricity regulatory commissions of 19 states and union territories have notified regulations for net metering/FIT mechanisms. The remaining states are being pursued to come out with their respective policies and regulations.

2.6.1.1 Fiscal and Financial Incentives for Rooftop PV

As the potential for rooftop SPV is sizeable, MNRE has made the following type of fiscal and financial incentives available for the specific benefit of both the commercial and industrial sectors.

- Custom duty concessions
- Excise duty exemptions
- Accelerated depreciation
- Fiscal and other concessions from the state governments

Importantly, MNRE has earmarked a CFA of 15% for the specific benefit of residential, institutional and social sectors.

Table 2.4   Estimated PV Rooftop Potential on Government Buildings

S. No.	Name of the Ministry	Indicative Potential (MW)
1.	Ministry of Agriculture	12
2.	Ministry of Chemicals and Fertilizers	401
3.	Ministry of Civil Aviation	620
4.	Ministry of Coal	53
5.	Ministry of Commerce and Industry	2
6.	Ministry of Consumer Affairs, Food and Public Distribution	2314
7.	Ministry of Culture	2
8.	Ministry of Defence	281
9.	Ministry of Food Processing Industries	22
10.	Ministry of Heavy Industries and Public Enterprises	271
11.	Ministry of Housing and Urban Poverty Alleviation	2
12.	Ministry of Human Resources Development	497
13.	Ministry of Micro, Small and Medium Enterprises	4
14.	Ministry of Petroleum and Natural Gas	1009
15.	Ministry of Railways	1369
16.	Ministry of Road Transport and Highways	0.4
17.	Ministry of Shipping	51
18.	Ministry of Steel	224
19.	Ministry of Textiles	5
20.	Ministry of Tourism	6
21.	Ministry of Youth Affairs and Sports	6

Source: Grid connected solar rooftop systems, Power Point presentation by Ministry of New and Renewable Energy, Government of India, www.mnre.gov.in/presentations [3].

Table 2.5 PV Installed Capacity Base across Diverse Sectors of Energy Economy

Sector	Installed by SECI (MW)	Installed by States (MW)	Total Installed Capacity (MW)
Commercial	10.90	17.22	28.91
Government	3.04	4.893	7.253
Institutions (schools and colleges)	2.19	5.131	8.346
Religious institutions	0.62	7.52	7.64
Residential	0.00	0.298	0.298
Total capacity (MW)	**18.35**	**35.532**	**54.187**

Source: MNRE.

2.6.2 A Few Selective Large-Capacity PV Rooftop Installations

Rooftop SPV deployment in increasing capacities has been witnessed both in India and overseas. Table 2.7 lists a few large rooftop installations with capacity shown in a descending order.

2.6.3 Case Study for an 80 kWp PV Rooftop System at Hero Honda

Hero Honda is the largest manufacturer of motorcycles in India, located at Dharuhera in Rewari district of Rajasthan. A SPV rooftop system of 80 kWp capacity was installed by Hero Honda after due diligence of a coordinated nature. The PV system thus installed qualified for a 30% capital subsidy as available in 2013 under the MNRE scheme for rooftops. The work that commenced after the receipt of subsidy in 6 months was completed in nearly 30 days. As per the available information, the PV system has been operating on an efficiency of 15% which implies a performance ratio of around 78%–80%. This system generated a total of 52,304.64 units over a period

Table 2.6 Status of Grid-Connected PV Rooftop Projects across Different States in India as of 30 June 2015

			Projects Sanctioned/Implementation under NCEF						Achievements		
Sl. No.	State/UTs	Projects Sanctioned under MNRE Scheme to SNAs/State Depts. (MWp)	NCEF I by SECI (26.6 MWp)	NCEF II by SECI (50 MWp)	NCEF III by MNRE to SNAs (54 MW)	NCEF IV by MNRE to MGAs[a] (52 + 52 MW)	NCEF V by SECI to Ware Houses (73 MW)	Total Sanctioned (MWp)	Under MNRE/ NCEF/S ECI (3–8)	Through Their Own Resources[b]	Total
1	2	3	4	5	6	7	8	9	10	11	12
1	Andhra Pradesh	5.5	2	3.5	4	0	0	15	2.46		2.455
2	Bihar	0	0	1	0	0	0	1	0.00		0.000
3	Chhattisgarh	0	2.05	0	5	0	0	7.05	1.75		1.750
4	Chandigarh	6.06	0.5	0	2	0	0	8.56	5.30		5.300
5	Delhi	0	2	2	8	25	0	37	4.32	3.87	8.190
6	Gujarat	5.75	0	2	0	0	0	7.75	0.00	9.75	9.750
7	Goa	0	0	0	2	0	0	2	0.00		0.000
8	Jharkhand	0	0	2	0	0	0	2	0.00		0.000
	J&K	0	0	0	0	0	0	0	0.00	1	1.000
9	Haryana	0	2	2	5	0	0	9	1.13		1.130

(Continued)

Table 2.6 (Continued) Status of Grid-Connected PV Rooftop Projects across Different States in India as of 30 June 2015

Sl. No.	State/UTs	Projects Sanctioned under MNRE Scheme to SNAs/State Depts. (MWp)	Projects Sanctioned/Implementation under NCEF					Total Sanctioned (MWp)	Achievements		
			NCEF I by SECI (26.6 MWp)	NCEF II by SECI (50 MWp)	NCEF III by MNRE to SNAs (54 MW)	NCEF IV by MNRE to MGAs[a] (52 + 52 MW)	NCEF V by SECI to Ware Houses (73 MW)		Under MNRE/ NCEF/S ECI (3–8)	Through Their Own Resources[b]	Total
1	2	3	4	5	6	7	8	9	10	11	12
10	Kerala	1.28	0	0	5	0	0	6.28	0.00		0.000
11	Karnataka	0	2	3	0	0	0	5	1.50	0.4	1.900
12	Madhya Pradesh	5	0.25	1	0	0	0	6.25	0.10		0.100
13	Maharashtra	0	2	5	0	0	0	7	0.67		0.670
14	Odisha	0	1	0	4	0	0	5	0.86		0.860
15	Punjab	5	0	2	0	0	0	7	0.00	7.52	7.520
16	Rajasthan	6	3.25	1	0	0	0	10.25	0.30		0.300
17	Tamil Nadu	6.74	5	5	5	0	0	21.74	6.60		6.600

(Continued)

Table 2.6 (*Continued*) Status of Grid-Connected PV Rooftop Projects across Different States in India as of 30 June 2015

			Projects Sanctioned/Implementation under NCEF						Achievements		
Sl. No.	State/UTs	Projects Sanctioned under MNRE Scheme to SNAs/State Depts. (MWp)	NCEF I by SECI (26.6 MWp)	NCEF II by SECI (50 MWp)	NCEF III by MNRE to SNAs (54 MW)	NCEF IV by MNRE to MGAs[a] (52 + 52 MW)	NCEF V by SECI to Ware Houses (73 MW)	Total Sanctioned (MWp)	Under MNRE/NCEF/SECI (3–8)	Through Their Own Resources[b]	Total
1	2	3	4	5	6	7	8	9	10	11	12
18	Tripura	0	1	0	0	0	0	1	0.00		0.000
19	Telangana	0	0	0	4	0	0	4	1.54		1.540
20	Uttarakhand	5	0	0	2	0	0	7	1.61	1.8	3.412
21	Uttar Pradesh	2	1.5	3	5	0	0	11.5	1.08		1.080
22	West Bengal	2.38	1	0	3	0	0	6.38	0.63		0.630
23	Ministry of Railways	0	2.5	0	0	50	0	52.5	0.00		0.000
24	Allocated to PSUs	0	0	0	0	19.79	0	19.79	0.00		0.000

(*Continued*)

Table 2.6 (Continued) Status of Grid-Connected PV Rooftop Projects across Different States in India as of 30 June 2015

Sl. No.	State/UTs	Projects Sanctioned under MNRE Scheme to SNAs/State Depts. (MWp)	Projects Sanctioned/Implementation under NCEF					Total Sanctioned (MWp)	Achievements		
			NCEF I by SECI (26.6 MWp)	NCEF II by SECI (50 MWp)	NCEF III by MNRE to SNAs (54 MW)	NCEF IV by MNRE to MGAs[a] (52 + 52 MW)	NCEF V by SECI to Ware Houses (73 MW)		Under MNRE/ NCEF/S ECI (3–8)	Through Their Own Resources[b]	Total
1	2	3	4	5	6	7	8	9	10	11	12
25	Pending allocation by SECI under NCEF	0	1.05	17.5	0	9.21	73	100.76	0.00		0.000
	Sub-total	50.71	29.1	50	54	104	73	360.81	29.847	24.34	54.187

Source: MNRE.

Notes: 20 MWp in Delhi Metro Rail Corporation.
5 MWp in NDMC Area (New Delhi).
1 MWp rooftop at Katra Railway Station by M o Railways.

[a] Agency-wise details given in Annexure.
[b] Own resources – 7.52 MWp in Dera Beas, Punjab and 1.80 MWp in IIT Roorkee, 300 kWp installed in Holy Family Hospital, Okhla, Delhi.

Table 2.7 Major PV Rooftop Capacities Installed across the World

Geographical Location	Country of Origin	Capacity Installed (MW)
Antwerp	Belgium	40 (in campus)
Radha Soami Satsang Beas, Amritsar	India	7.52 (single roof)
Bay Resort DLR Group – Mandalay Convention Centre, Las Vegas, NV	United States	6.4
Constellation Energy – Toys R US, Flanders, NJ	United States	5.38
Southern California Edison, Fontana, CA	United States	2.0
Boeing 787 Assembly Building, North Charleston, SC	United States	2.6

Source: MNRE.

of 6.7 months, that is from 22 September 2013 to 4 April 2014. Buoyed by the success of the solar plant, the company is now embarking on a plan to install more plants especially of a megawatt scale. Table 2.8 highlights the salient features of the 80 kWp PV power plant.

2.7 Status of Large-Capacity Grid-Connected Power Generation

As per the latest statistics by MNRE, India's total grid-connected solar-generation capacity stood at 3883.5 MW (as of 29 May 2015). This capacity has since then increased to 4200 MW as on 31 August 2015. Seven states had more than 100 MW installed capacity and about 53% of the installed capacities were from state schemes, 32% were from MNRE projects and the remaining

Table 2.8   Key Attributes of a PV Rooftop Facility of a Major Automaker

Module Technology	Crystalline silicon
Module Make	Hindustan High Vacuum, Bangalore
Inverter Type	Centralized
Inverter Make	SMA, United States
Utility Supplier	Haryana State Electricity Board
Contracted Demand from Utility Supplier	4000 kVA
Captive Power Plant-I Captive Power Plant-II	6 MW – 2 units of 3 MW, on-site natural gas plant 1 1 MW – 5 units of 2 MW and 1 unit of 1 MW on-site diesel-powered plant
Total Project Cost	Rs. 80 lacs

Source: Employment generation potential of a solar PV rooftop system at Hero Honda Motors, Issue paper, August 2015, www.mnre. gov.in [4].

through the REC route. Table 2.9 gives a state-wise break-up of the PV power capacities installed in India till May 2015.

According to the latest statistics by the MNRE, a total of 1104 MW of solar power projects were commissioned during the period from April 2014 to March 2015. A total of 14 states added solar power generation capacity, and the state of Rajasthan (229 MW) had the highest installed capacity during the year. Three other states (Madhya Pradesh, Punjab and Andhra Pradesh) installed more than 100 MW each.

Figure 2.2 provides the state-wise break-up of the PV installed capacity.

2.7.1 Capacity Addition under Various Schemes

About 74% of the total capacity addition took place through MNRE schemes (Phase 2, Batch 1 of JNNSM) and state policies. A total of 454 MW of projects were commissioned under MNRE schemes and 360 MW of projects under state

Table 2.9   PV Installed Capacity across Different States in India under Various Schemes

State-Wise Installed Capacity of Solar PV Projects under Various Scheme as on 29 May 2015

Sr. No.	State/UT	Total MNRE Projects MW	State Schemes MW	Total MW	Total MW	% Share
1	Rajasthan	789.1	65.0	193.0	1047.1	27.0
2	Gujarat	20.0	974.1	6.0	1000.1	25.8
3	Madhya Pradesh	185.3	297.6	80.8	563.6	14.5
4	Maharashtra	57.0	185.4	121.3	363.7	9.4
5	Andhra Pradesh	94.8	115.0	37.7	247.5	6.4
6	Punjab	10.5	177.3	7.5	195.3	5.0
7	Tamil Nadu	16.0	33.8	98.2	148.0	3.8
8	Karnataka	5.0	73.2	0.0	78.2	2.0
9	Uttar Pradesh	12.0	59.3	0.0	71.3	1.8
10	Telangana	0.0	39.4	23.4	62.8	1.6
11	Orissa	12.0	15.4	4.5	31.9	0.8
12	Jharkhand	16.0	0.0	0.0	16.0	0.4
13	Haryana	7.8	5.0	0.0	12.8	0.3

(Continued)

Table 2.9 (Continued) PV Installed Capacity across Different States in India under Various Schemes

| | State-Wise Installed Capacity of Solar PV Projects under Various Scheme as on 29 May 2015 | | | | | |
| | Total MNRE Projects | State Schemes | | Total | | |
Sr. No.	State/UT	MW	MW	MW	MW	% Share
14	Chhattisgarh	4.0	1.7	4.6	10.3	0.3
15	West Bengal	2.1	5.2	0.0	7.2	0.2
16	Delhi	0.3	4.2	2.1	6.7	0.2
17	Andaman and Nicobar	0.1	5.0	0.0	5.1	0.1
18	Uttarakhand	5.0	0.0	0.0	5.0	0.1
19	Tripura	0.0	0.0	5.0	4.5	0.1
20	Chandigarh	4.5	0.0	0.0	0.8	0.1
21	Lakshadweep	0.8	0.0	0.0	0.0	0.02
22	Arunachal Pradesh	0.03	0.0	0.0	0.0	0.001
23	Kerala	0.03	0.0	0.0	0.0	0.001
24	Puducherry	0.03	0.0	0.0	0.0	0.001
25	Others	0.8	0.0	0.0	0.8	0.02
	Total	**1243.0**	**2056.4**	**584.1**	**3883.5**	**100**

Source: MNRE.

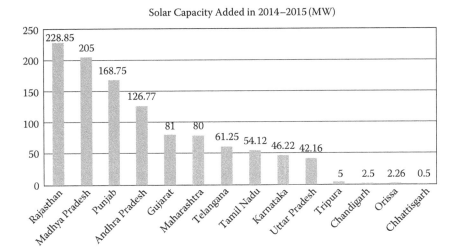

Solar Capacity Added in 2014–2015 (MW)

Figure 2.2   State-wise break-up of PV installed capacity.

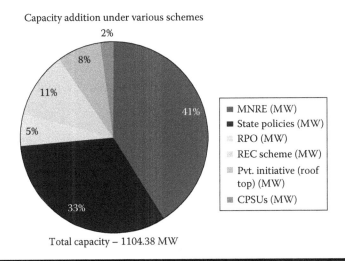

Capacity addition under various schemes

Total capacity – 1104.38 MW

■ MNRE (MW)
■ State policies (MW)
▨ RPO (MW)
▨ REC scheme (MW)
▨ Pvt. initiative (roof top) (MW)
■ CPSUs (MW)

Figure 2.3   Capacity addition vis-à-vis different schemes. (From RE-Invest 2015, A special report of the Ministry of New and Renewable Energy, Government of India [5].)

schemes. Amongst state policies, Punjab led the capacity addition with 167 MW, which is close to 50% of the total capacity addition under various state schemes. Uttar Pradesh, Andhra Pradesh, Telangana and Karnataka also added significant capacity through their respective state solar policies (Figure 2.3).

According to the official estimates of MNRE, this is the first time that more than 1000 MW have been added in a financial year (April–March). With this, the total grid-interactive solar capacity reached 3736 MW as of 31 March 2015. With several states coming up with various policies, it is very likely that the cumulative solar capacity will cross the 5 GW mark during the financial year 2015–2016.

References

1. Solar PV Rooftop Systems in India 2015, Shakti Foundation Report, www.shaktifoundation.in.
2. Technical standards for connectivity of the distributed generation resources, report by Central Electricity Authority, Indian Ministry of Power, 2013, www.cea.nic.in.
3. Grid connected solar rooftop systems, Power Point presentation by Ministry of New and Renewable Energy, Government of India, www.mnre.gov.in.
4. Employment generation potential of a solar PV rooftop system at Hero Honda Motors, Issue paper, August 2015, Inhouse publication of Hero Motors Corporation.
5. RE-Invest 2015, A special report of the Ministry of New and Renewable Energy, Government of India.

Chapter 3

Solar Radiation Availability on Earth

3.1 Introduction

The Sun is a gigantic element of the solar system. In actual terms, it is a sphere comprising, essentially, hot gaseous matter. The diameter of the Sun is equal to 1.39×10^9 m and is at a distance of 1.5×10^{11} from the Earth. The Sun has an effective black body temperature Ts of about 5777 K and it accounts for about 99.86% mass of the solar system. The mean distance of the Sun from the Earth is nearly 149.6 million km and sunlight covers this distance in about 8 min and 19 s. This distance changes across the year ranging from a minimum of 147.1 million km on the perihelion to a maximum of 152.1 million km on the aphelion. In terms of the gaseous composition, the Sun has hydrogen equivalent to around 74% of its mass or 92% of its volume, followed up by helium (about 24% of mass, 7% of volume). In addition, it has trace quantities of other elements such as iron, nickel, oxygen, silicon, sulphur, magnesium, carbon, neon, calcium and chromium. Importantly, many fusion reactions occur on the Sun's surface with the most prominent of these being hydrogen (i.e. four protons).

These react to form helium; the mass of the helium nucleus is lower than that of four protons.

The rate of energy released from the Sun is 3.8×10^{23} kW, out of which, the Earth intercepts only a tiny fraction. This amounts to around 1.7×10^{14} kW. However, even such a tiny quantum is several thousand times more than the existing consumption rate of energy on the Earth. Due to this key attribute, the Sun can be considered as a source of infinite energy. However, solar radiation is a dilute source of energy with a maximum flow density of about 1.3 kW/m². This is generally seen as a low value from the key consideration of a technological utilization. Moreover, owing to the geometry of Earth–Sun movements, there are significant changes in the amount of solar radiation received at any given location. The largest change is observed at the poles. Such variations occur both on a diurnal (i.e. during the day) and seasonal basis. Further, the presence of clouds and dust in the atmosphere decreases the availability of solar energy. Therefore, it is essential to have accurate information about the quality and quantity of solar resources at a specific location. This is an important requirement for the optimal designing of the equipment.

3.2 About the Earth

The Earth is the largest of the terrestrial planets within the solar system in terms of mass, density and diameter. The Earth has a diameter of nearly 12,800 km and completes one rotation about its own axis every 24 h. It also completes a revolution around the Sun in a period of 365.25 days. The surface area of the Earth is 51,00,72,000 km², out of which nearly 29.2% is land area. The circumference of the Earth at equator is around 25,000 miles. The area covered by water is about 70.8%.

The orbital path is equivalent to 450 million km and the time taken is equal to 365.25 mean solar days. Importantly, the speed of rotation is equal to 30 km/s. The equatorial plane is tipped

23.5° from the ecliptic plane. As the Earth revolves around the Sun, this orientation produces a varying solar declination.

3.2.1 Key Outcome of Earth's Rotation

The Earth's rotation around the Sun results in the following:

- Days and nights
- Seasons
- Sunshine hours per day
- Variations in solar radiation

Peak Sun hours is an equivalent measure of total solar radiation in a day.

3.2.2 Energy Balance

It makes good sense to take a close look at the energy balance of Earth. The global energy balance is represented in Watts per square metre. Figure 3.1 showcases the attributing factors

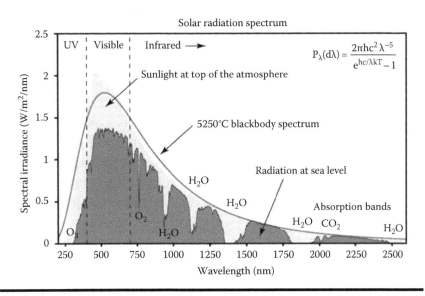

Figure 3.1 Attributing factors towards absorption/scattering of the radiation spectrum. (From www.wikipedia.org. [1])

which are responsible for absorption/reflection/emission from various sources. The Earth reflects nearly one-third of the radiation which falls upon it. This is also referred to as the Earth's albedo in common terms. The Earth keeps spinning about its axis on a continuous basis and its axis is inclined at an angle of about 23.45°.

Additionally, this inclined position of the Earth along with the Earth's daily rotation and annual revolution accounts for the following:

■ Distribution of solar radiation over the Earth's surface.
■ Changing length of the hours of daylight, darkness besides the changing surface.

3.3 Radiation Emissions

It is widely perceived that matter emits electromagnetic radiation which is usually referred to as thermal or heat radiation. The fact is that all substances, solid structures as well as liquid and gases, above the temperature of absolute zero emit energy in the form of electromagnetic waves. The Sun emits energy in the form of electromagnetic radiation distributed over a wide range of wavelengths. The wavelengths less than around 0.4 µm are known as ultraviolet (UV), that is, shortwave, whereas wavelengths longer than about 0.7 µm are infrared, that is, long wavelengths. The visible radiation range is located between the UV and IR rays. The maximum spectral intensity takes place at about 0.48 µm wavelength in the green part of the visible spectrum. In terms of energy presence, it is as per Table 3.1.

In total, about 99% of the energy of solar radiation is contained in the wavelength band of 0.15–4.0 µm which includes UV, visible and IR regions. The radiations which are released from the Sun during their passage through the Earth's

Table 3.1 Percentage Contribution to Total Energy by Different Wavelength Spectra

Ultraviolet region (UV)	Equivalent to about 8.73% of the total energy
Visible region (IR)	Equivalent to about 38.15% of the total energy
Infrared region (VR)	Equivalent to about 53.12% of the total energy

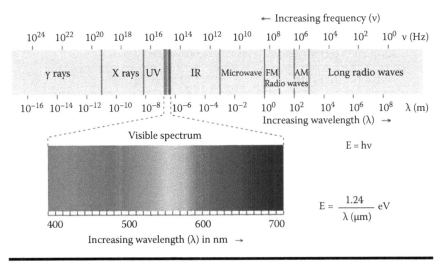

Figure 3.2 Electromagnetic spectrum at a glance. (From www. wikipedia.org. [2])

atmosphere are subjected to the mechanism of attenuation. Figure 3.2 demonstrates the electromagnetic spectrum.

This is mainly due to the atmospheric absorption and scattering. Importantly, the Earth's atmosphere has a unique property of absorbing UV and far IR radiation, which allows only 0.29 and 0.23 μm of radiation. Table 3.2 shows the physical properties of the Sun and Earth.

Solar irradiance is solar power per unit area. The inverse square law states that irradiance is reduced in proportion to the inverse square of the distance from the source.

Table 3.2   Combined Physical Properties of Sun and Earth

Surface area of the Sun	6.093×10^{12} km^2
Mass of the Sun	1.989×10^{33} g
Volume of the Sun	1.4122×10^{33} cm^3
Temperature of solar corona	10,00,000 K
Rate of Sun's radiation	6.5×10^{10}/erg/cm^2
Period of rotation of the Sun at the equator	24.65 days
Value of solar constant	1370 W/m^2
Rate of energy production	3.90×10^{16} W
Mass of the Earth	5.517×10^{27} g
Mean radius of the Earth	6371 km
Surface area of the Earth	5.101×10^{14} m^2

Source: Garg, H.P., *Treatise on Solar Energy, Vol. 1: Fundamentals of Solar Energy* [2].

3.4 Solar Radiation Terminology

Solar constant: The orbit of the Earth is elliptical in shape, owing to which the distance between the Sun and the Earth differs by about 1.7%. At a mean Earth–Sun distance of about 1.495×10^{11} m, the Sun subtends an angle of 32°. The Solar Constant (G_{sc}) is the energy received from the Sun per unit time on a unit area of surface perpendicular to the direction of propagation of the radiation at the mean Earth–Sun distance. World Radiation Centre has assigned a value of 1367 W/m^2.

Albedo: The albedo of an object is the extent to which it reflects diffused light from the Sun. Hence, it is a more specific form of the term 'reflectivity'. Simply put, albedo (a dimensionless quantity) is defined as the ratio of diffusely reflected to incident electromagnetic radiation.

Zenith: It is the angle between the vertical and the line of the Sun, that is the angle of incidence of direct solar radiation or a horizontal surface.

Solar time: This is the time based on the apparent angular motion of the Sun across the sky and does not coincide with the local clock time. For example, the solar noon is the time at which the Sun crosses the meridian of the observer which need not be 12 noon as per the local time.

Air mass: It is the path length of radiation through the atmosphere, considering the vertical path at sea level as unity. At sea level, m = 1, when the Sun is at the zenith (that is directly overhead) and m = 2 for a zenith angle θ_z of 60°. Air mass is a representation of the amount of the atmosphere radiation that must pass through to reach the Earth's surface. There are two main air mass terms – AM 0, which above the Earth's atmosphere is equivalent to 1366 W/m², and air mass one (AM1), which is the radiation that crosses the atmosphere directly from over-head (i.e. zenith).

Beam radiation (*direct*): The portion of solar radiation that strikes the Earth without any alteration in direc-tion is called as beam radiation or in more common terms as direct radiation. It is the part of sunlight that comes directly from the Sun. This would exclude diffuse radiation such as that which would come through on a cloudy day. It also gives an indication about the clear-ness of the sky.

Diffuse radiation: It is the solar radiation which is received by the Earth after a change in its direction. This occurs mainly due to the scattering in the atmosphere. This is basically the solar radiation reaching the Earth's surface after having been scattered from the direct solar beam by the molecules in the atmosphere. This type of radia-tion is also known as skylight, diffuse sunlight or even sky radiation and is the reason for changes in the colour of the sky. Further, out of the total light removed from the direct solar beam by scattering in the atmosphere (i.e. nearly 25% of the incident radiation when the Sun is high in the sky, depending on the amount of dust and haze in the atmosphere). Almost two-thirds finally reach the Earth in the form of diffuse solar radiation.

Total solar radiation: It is the direct addition of the beam and diffuse components of solar radiation. Total solar radiation on a horizontal surface is generally known as global radiation. This is also known as global horizontal radiation and is the sum of direct, diffuse and ground-reflected radiation. However, ground-reflected radiation is generally minuscule compared to both the direct and diffuse radiation types. Thus, global radiation is assumed to be the sum of direct and diffuse radiation for all practical purposes.

Irradiance: It is the rate at which the radiant energy is incident on a unit area of surface. This is usually marked by W/m^2 and applicable for beam, diffuse or spectral distribution.

Insolation: Insolation is basically a measure of solar radiation received on a given surface area in a given time. It is commonly expressed as average irradiance in watts per square metre (W/m^2) per day. In the case of solar photovoltaic (PV), it is generally measured in $kWh/kWp·y$, that is kilowatt hours per year per kilowatt peak rating. The incident solar energy radiation or irradiation is also termed as insolation. In terms of notations, H is the insolation for the day and I is the insolation for a specific time period, usually 1 h. Both H and I are expressed as $W·h/m^2/day$ and $W·h/m^2/h$. When the values are measured on an hourly basis, I is numerically equal to G.

3.4.1 Solar Radiation on Different Sources

Solar radiation entering the atmosphere comprises direct, diffuse and albedo radiation. There exists a geometric

relationship between a plane of any particular orientation relative to the Earth at any time regardless of the fact
whether the plane is fizzed or moving relative to the Earth.
The incoming beam radiation that is the position of the Sun
relative to that plane can be described in terms of different
angles as follows.

Latitude: It is the angle made by the radial line joining the
centres of the Sun and Earth with its projection on the
equatorial plane. Latitude varies between −90° and +90°
and is positive for the northern hemisphere.

Solar declination: The Earth's axis of rotation is inclined at
an angle of 23.45° to the axis of its orbit around the Sun.
This tilt causes seasonal variations at any location. The
angle between the Earth–Sun line (i.e. through their centres) and the plane through the equator is called the solar
declination. It changes between −23.45° on 23 December
and +23.45° on 21 June. Further declinations towards the
north of equator are positive whereas those towards the
south are negative.

Slope: It is the angle between the plane of surface concerned and the horizontal plane.

Surface azimuth angle: It is the angle made in the horizontal plane between the line due south and the projection
of the normal to the surface on the horizontal plane. As
per the convention, due south is taken as zero, east of
south is positive and west of south is negative.

Hour angle: It is the angular displacement of the Sun, east
or west of the local meridian, due to the rotation of the
Earth on its axis at 150 h⁻¹; morning is negative and after-
noon is positive.

Angle of incidence: It is the angle between a ray incident on
a surface and the normal to that surface.

Zenith angle: It is the angle between the vertical plane and
the line to the Sun, that is the angle of incidence of the
beam radiation on a horizontal surface.

Solar altitude angle: It is the angle between the horizontal plane and the line of the Sun, that is the complement of the zenith angle.

Solar azimuth angle: It is the angular displacement from the south of the projection of the beam radiation on a horizontal plane. The displacements of the east of south are negative and those of the west of south are positive.

3.5 Sun–Earth Geometric Relationship

As mentioned before, the distance between the Earth and the Sun changes from time to time due to the Earth's elliptical orbit. The minimum distance is 147.1×10^6 km at the winter solstice on 21 December. This point in the orbit is known as perihelion. The maximum distance is 152.1×10^6 km at the summer solstice on 21 June. This point in the orbit is known as aphelion. The difference between the maximum and minimum distances is around 3.4%. The angles at which the Sun's rays impinge on the Earth in different seasons and location of tropics is shown in Figure 3.3.

As the Sun is far away from the Earth, all its rays may be deemed as being parallel to one another when they reach the Earth. The Earth rotates on its axis and it also revolves around the Sun in an elliptical orbit. The area swept by the Earth–Sun radius remains constant much in accordance with the Kepler's law of planetary motion. In order to make calculations linked with the flux density of solar radiation on flat and inclined surfaces, it must be possible to evaluate the angle of the Sun from any position on the Earth at any time of the day. Such a calculation involves computing the angle between a vector normal to a specified point on the Earth and a projection from the Sun. However, this type of computation assumes some complexity as the Earth rotates on its polar axis once a day, and the Earth revolves around the Sun once a year. Further, to determine the Suns elevation with respect to the horizon

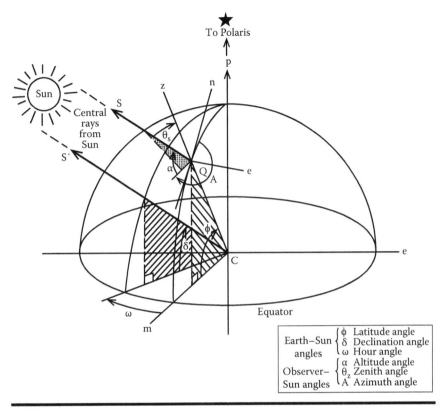

Figure 3.3 Schematic representation of different angles related to incidence of sunrays on Earth.

and its azimuth angles, it is essential to take into consideration the concepts evolved from the three-dimensional geometry. Following are the few vital coordinates which must be known in this connection:

■ Time of the day
■ Longitude
■ Latitude
■ Declination angle of the Earth

Figure 3.4 [3] gives a schematic representation of the solar radiation geometry.

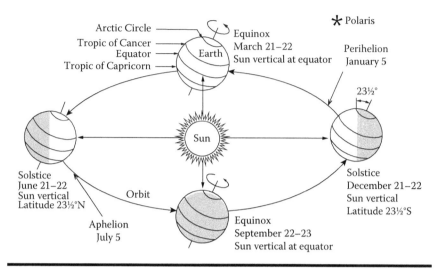

Figure 3.4 Solar radiation geometry depiction.

3.5.1 Sun's Trajectories during Different Seasons

The rotation of the Earth around the Sun gives rise to the seasons. This is because the Earth's axis of rotation is tilted at 23.45° and is not perpendicular to the plane of its orbit. Trajectories of the Sun as seen from a point on the Earth change from month to month as shown in Figure 3.5. During the northern summer, it is tilted away from the Sun. This tilt causes the Sun to appear higher in the sky in summer, thus causing more hours of daylight and more intense direct sunlight or hotter conditions on the surface of the Earth. In contrast, during winter, the Sun's rays hit the Earth at a shallow angle. As these rays are more spread out, the energy that hits any given spot gets minimized. At the start of April and September, the Earth is at a mean distance of 149.6×10^6 km from the Sun.

The Sun's visible diameter also changes as the Earth moves around its orbit. In January, the angle made by the diameter of the disc is at a maximum of 32.36, and in July, it is at a minimum of 31.32. As and when the Earth is at its mean distance from the Sun, this angle is about 32°. The changes in distance across the year lead to changes in the radiation flux reaching

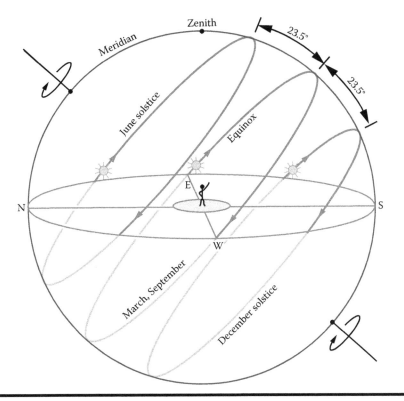

Figure 3.5 Sun-based trajectories on a year-round basis.

the Earth from the Sun. The radiation flux changes inversely with the square of the distance.

3.5.2 Key Factors on Which Insolation Depends

Solar insolation is a key determinant on which the power availability from a PV array depends. In fact, insolation has a strong bearing on three factors. The basic unit of solar energy is watts per square metre. Out of these, the first factor is solar constant.

Solar constant: It is the amount of solar energy at normal incidence outside the atmosphere.

Elevation: It is the elevation of the Sun in the sky.

Quantity of radiation (reflected): This is the amount of radiation which is reflected back into the space by the Earth's atmosphere.

3.5.3 Quantity of Radiation (Absorbed)

It is the amount of radiation which is absorbed by the atmosphere in addition to the amount of radiation reflected by the lower boundary of the Earth. In turn, the amount of usable radiation available at a site is strongly dependent on the factors mentioned in Table 3.3.

Solar radiation is available on the Earth's surface in varying amounts throughout the day. The quality of solar radiation and the amount of sunlight available at any given location on the Earth are much in accordance with the factors highlighted in Table 3.4.

The amount of solar radiation can be equivalent to 1000 W/m^2 when the sunrays strike the Earth's surface at 90°. This is quite likely to happen at solar noon. Solar irradiance above the Earth's atmosphere typically varies between 1325 and 1412 W/m^2. As such, solar constant is the average of this range, that is 1367 W/m^2. As has been observed, solar irradiance at the Earth's surface is much lower as the solar rays lose their strength during their passage through the atmosphere. Table 3.5 gives a brief representation of power densities of published standards.

Table 3.3   Dependence of Usable Solar Radiation on Different Factors

Latitude	The location of a site on the planet affects the angle of the incident sunlight and in turn the amount of solar radiation available to us
Local conditions	These mainly comprise trees or obstruction which might shade the solar arrays Other conditions include dirt, dust or local airborne pollutants such as kerosene/diesel residues and airflow
Climatic characteristics	Prominent climatic conditions which can lower down the concentration of the solar radiation may mainly include clouds and snow cover

Table 3.4 Availability of the Sunlight in Accordance with Different Factors

Shape	The round shape of the Earth makes sunlight strike the planet at different angles
Earth's path around the Sun	The orbit of the Earth enables sunlight to hit the Earth' surface at different angles and intensities, hourly, daily, monthly, seasonally and annually
Elliptical orbit	The Earth revolves around the Sun in an elliptical orbit. This brings the Earth nearer to and farther from the Sun
Axis	The North Pole points away from the Sun during winter in the northern hemisphere, while the South Pole points towards the Sun This results in differences in the available energy between the North Pole and the South Pole Sunrays strike the Earth at an oblique angle beginning at 0° with the Sun just above the horizon The time the Sun is directly overhead, that is at solar noon on the equator, the rays strike the Earth perpendicularly, that is at 90°

Table 3.5 Power Density Figures at a Glance

Solar Condition	Applicable Standard	Power Density (W/m²)		
		Total	250–2500 nm	250–1100 nm
	WMO Spectrum	1367		
AM 0	ASTM E 490	1353	1302.6	1006.9
AM 1	CIE Publication 85, Table 2		969.7	779.4
AM 1.5 D	ASTM E 891		969.7	779.4
AM 1.5 G	ASTM E 892	963.8	951.5	768.6
AM 1.5 G	CEI/IEC*904-3	1000	987.2	797.5

Source: Commission of Electrotechnique International/International Electrotechnical Commission. CEI, Commission Electro technique Internationale; IEC, International Electrotechnical Commission [4].

3.5.4 Radiation on a Tilted Surface

In reality, the solar modules are not positioned in a horizontal position. Instead, the modules are installed at an angle with the horizontal surface to maximize the collection. As such, it becomes logical to estimate the solar radiation on the tilted surfaces for an all-important purpose of designing or evaluating the field performance of solar systems. The incident solar radiation on a tilted surface comprises the following:

- Beam radiation
- Three components of diffuse radiation from the sky
- Radiation reflected from such surfaces as are seen by the tilted surface

3.5.5 Concept of In-Plane Radiation

In-plane radiation is widely regarded as the most relevant term for solar PV power generation. It is made up of the global (direct and diffused radiation) radiation. Orientation together with the tilt angle of the plane needs to be optimized so as to maximize the in-plane radiation. This depends on the location on the Earth itself. Further, one may track the Sun so as to enhance the in-plane radiation either by single-axis or by double-axis technique. Solar window is the area of sky which contains all possible locations of the Sun across the year for a particular location. Array orientation can be described using azimuth and tilt angles. Importantly, both these parameters are vital determinants for the estimation of power output. It is possible to optimize energy production at certain times of the year by adjusting the array tilt angle. Several opinions point to having tilt angle equal to the latitude of a given location. However, the latitude tilt of the solar modules which face south (i.e. with azimuth = 0°) is not necessarily the best optimization option available to a solar designer. Further, several geographical parameters and nearby bodies such as water bodies and hills may exert some effect on the in-plane solar radiation. In turn, it means a reduced photovoltaic yield too.

3.5.5.1 Case-Specific Example

There is a visible difference when it comes to gain in solar radiation with a fixed tilt system and that with a seasonal tilt system. Field-related observations point to a gainful value of around 8.46% for a fixed tilt system as against 12.66% in case of a system with a seasonal system. This involves twice a year seasonal type adjustment. Again annual gain in solar radiation with zero-tilt single-axis tracking as measured for a particular site is equal to 24.53% as against 31.19% for a system with latitude single-axis tracking system.

3.5.5.2 Empirical Estimation of Solar Radiation

The estimation of monthly averaged global radiation on a horizontal surface is dependent on the following few parameters:

H_{ga} = monthly averaged daily global radiation on a horizontal surface

H_{oa} = monthly averaged extra-terrestrial solar radiation at horizontal surface (at the top of atmosphere)

S_a and S_{maxa} = monthly averaged daily sunshine hours and maximum possible daily sunshine hours (the day length) at a given location

The standard equation used is as follows:

$$H_{ga}/H_{oa} = a + b(S_a/S_{maxa})$$

a and b are constants.

Table 3.6 shows the values of constants a and b across several important stations worldwide.

Table 3.6   Solar Constants for Various Locations Worldwide

Location	Country of Origin	a	b
New Delhi	India	0.25	0.57
Nagpur	India	0.27	0.56
Ahmedabad	India	0.28	0.48
Bengaluru	India	0.18	0.64
Pune	India	0.31	0.43
Atlanta, Georgia	USA	0.38	0.26
Hamburg	Germany	0.22	0.57
Miami, Florida	USA	0.42	0.22
Nice	France	0.17	0.63
Rafah	Egypt	0.36	0.35

Source: Solar constant data tables consolidated from multiple sources [5].

3.6 Radiation Measurement Techniques

Satellites do not measure radiation as such. Instead they are equipped with on-board cameras that click the pictures. Obviously, such pictures need to be interpreted minutely and the camera pixel holds the key to actual resolution of the picture. Thus, highly sophisticated methodologies are developed to obtain solar radiation data from satellite pictures. Table 3.7 highlights the popular type of available data sources for solar radiation. This includes the source of database in addition to mentioning the value, period and the variables determined.

Solar radiation mainly constitutes the direct (beam) radiation, which comes straight from the Sun, and diffused radiation, which comes from everywhere and not directly from the Sun. Global radiation is the sum of both the direct and diffused types of radiation. The author (i.e. Suneel Deambi) has extensively worked with the field use of pyranometer during a MNRE supported

Table 3.7 Prominent Available Data Sources for Solar Radiation Worldwide

Database Source	Region	Values	Source	Period	Variables	Availability
Meteonorm	Worldwide	Monthly/ hourly	1700 Terr. Stations (Interpolations)	1981–2000	GHI, DNI, DHI, in-plane, temperature, wind, etc.	Software (free/ paid)
SSE-NASA	Worldwide	Monthly	Satellite 1 in./1 in. cell	1983–1993	GHI, DNI, DHI, in-plane, temperature, wind, etc.	Web (free)
WRDC	Worldwide	Hourly, daily, monthly	1195 stations	1964–1993	GHI, DHI and sunshine	Web (free)
RETScreen	Worldwide	Monthly	Compil. 20 sources including WRDC-NASA	1961–1990	GHI, DNI, DHI, in-plane, temperature, wind, etc.	Software (free)
Indian Meteorological Department	India only	Hourly, daily, monthly	23 Stations only		Solar radiation and weather parameters	Payment basis

(Continued)

Table 3.7 (Continued) Prominent Available Data Sources for Solar Radiation Worldwide

Database Source	Region	Values	Source	Period	Variables	Availability
3 Tier	Worldwide	Monthly	3 km × 3 km (satellites)	December 1998–present	GHI, DNI and DHI	Web (paid)
SolarGIS	Europe, Africa, West Asia, West Australia and Brazil	Monthly	5 km × 5 km (Meteostat)	1998–up-to-date	GHI, DNI, Temp, RH, WS and WD	Web (paid)
NREL NEW	Worldwide	Monthly and hourly	10 km × 10 km Satellite (MISR, MERIS and MODIS)	2002–2008	GHI, DNI and DHI	Web (free)

project – 'Performance evaluation of PV power packs in rural areas of northern India' – during his tenure with the Energy and Resources Institute (TERI) between 1990 and 1994. The purpose was to measure the PV module parameters in relation to the changing values of solar insolation across the day. Pyranometer is an instrument used to measure the total or global solar radiation. It is also capable of measuring the diffuse radiation when it is provided with a shading ring. The instrument usually measures solar radiation in a wavelength range of 0.13–3.0 µm. Key constituents of a commonly used pyranometer are as follows:

■ Detector or sensor which is duly shielded by a glass dome
■ Instrument body with a spirit level together with adjustable levelling screws and a desiccant chamber
■ Radiation shield so as to protect the instrument case from direct sunlight
■ Electrical connector for the output signal

3.6.1 Different Types of Sensors

Solar radiation sensors are generally of two main types – photosensors and thermal sensors. Silicon-made solar cells serve as a photosensor. However, these types of sensors respond selectively vis-à-vis the wavelength of incident solar radiation. Because of this, their use is limited to the measurement of special wavelengths or short spectral bands. To remedy this constraint, thermal sensors are usually used. This ensures an accurate measurement over the full wavelength range. The primary function of a thermal sensor is to convert the heating effect of solar radiation into electrical signals.

The common types of pyranometers utilize the thermoelectric principle, where thermocouples are arranged in series (i.e. thermopile). A black and white pyranometer has a radial wire–wound thermopile as a detector. It consists of three segments coated with black lacquer, which form the hot junction of the thermocouple. The remaining three segments are coated with white barium sulphate. As it is well known, white colour acts as a good reflector of solar radiation. Hence the white-coloured segments act as cold junction and possess a large thermal inertia to overcome thermal fluctuations. In addition, the instrument has an embedded circuit, that is temperature compensation circuit. This reduces the effect of ambient temperature fluctuations. Further, it is placed with two hemispherical domes, where the inner concentric dome shields the sensors from transient convection effects. The following factors should be taken into account for ensuring reliable measurements of incident solar radiation:

■ Sensitivity of the order of 5 $MV/cal/cm^2/min$
■ Response changes due to fluctuations in ambient temperature (±0.5% from −20°C to 40°C)
■ Response time of 1 s

It is important to carry out a periodic calibration by either pyrheliometer or calibrated pyranometer as a reference tool.

3.6.2 Different Types of Pyranometers

Pyranometer with shaded ring: This device is inexpensive and is used for the specific measurement of diffuse radiation. However, a key constraint is that the measurements made need to be corrected. This is because the shading ring partially obstructs the passage of solar radiation. Figure 3.6 shows a pictorial representation of a pyranometer.

Shaded pyranometer: This device is used to measure diffuse solar radiation. However, it needs a shading device whose purpose is to shade the sensor from direct solar beam.

Pyranometer with shading disc: This device along with a shading disc is also used to measure diffuse solar radiation. In this case, the measured data do not need any correction. This is because the disc obstructs only the path of direct radiation.

Figure 3.6   View of a pyranometer.

3.6.3 Measurement of Direct Radiation

Pyrheliometer: In common terms, direct radiation is also referred to as beam radiation. The instrument which is generally used to measure direct radiation is known as pyrheliometer. As is obvious, this instrument must be devoid of diffuse component. This is attained by using a collimator tube over the Sun in addition to continuously tracking the Sun. The Angstrom electrical compensation pyrheliometer is one of the most reliable instruments widely used to measure direct radiation. It is made of two strips of manganin foil which is coated on one side with Parsons black lacquer. Figure 3.7 shows a basic representation of a pyrheliometer.

The strips are placed side by side, and a thermojunction is attached at the back of each strip. Further, this is placed in a holder, which is placed at the base of a cylindrical metal tube better known as the collimator. The collimating tube which is blackened from inside normally offers a 5 degrees field of view. This tube is filled with dry air at atmospheric pressure with its viewing end sealed by a removable, 1 mm thick, crystal quartz window. Further, an azimuth elevation mechanism is provided to help in directing the cylindrical tube at the Sun.

Figure 3.7   Pictorial representation of a pyrheliometer.

In addition, it also has a reversible shutter fixed at the front end of the tube. This allows one of the two strips to remain shaded from the Sun. The other strip is fully exposed to the direct solar radiation.

Underlying operating principle: In practical terms, the strip which is exposed gets heated by the direct radiation from the Sun. As a result, the temperature goes up. In contrast, the strip which remains shaded is heated electrically to the same temperature as the exposed strip. This is done to ensure that the rate of energy absorbed by the exposed strip is thermoelectrically compensated by the rate of electrical energy supplied to the shaded strip. The uniformity of temperatures of these two strips is ascertained by the thermojunctions attached to the reverse side of the strips via a sensitive null detector.

3.6.3.1 *Broad Classifications of Pyrheliometers*

World Meteorological Organization (WMO) set up in 1950 is an apex body of the United Nations drawing membership from as many as 191 states and territories. It is an acknowledged specialized body for meteorology, that is weather and climate.

WMO has classified pyrheliometers much in accordance with their accuracy as well as the precision of their auxiliary equipment. The accuracy depends on the following parameters:

- Sensitivity of the instrument
- Stability of the correction factor
- Maximum error due to changes in the ambient temperature
- Errors due to spectral response of the receiver
- Non-linearity of the response
- Opening angle
- Time constant of the system
- Resultant effect of the auxiliary equipment

Based on this, pyrheliometers are broadly classified as follows:

- Absolute pyrheliometer
- Reference standard pyrheliometer
- First-class pyrheliometer
- Second-class pyrheliometer

It is pertinent to mention here that the World Radiation Centre (WRC) at Davos maintains the absolute pyrheliometers against which standard pyrheliometers are usually calibrated. These reference standard pyrheliometers are subsequently put to use for the desired calibration of the remaining few types, that is, first-class and second-class type Pyrheliometers.

3.6.4 Sunshine Recorder

This device, as the name suggests, is generally used to measure the available amount of sunshine. Actually it measures the number of insolation hours only – the sunshine that is directly present. The sunshine duration is defined as the time during which the Sun's disc is not hidden by clouds. Further, the quantity measured by this device is the time in which direct solar radiation has enough intensity to activate the recorder. In actual terms, the sunshine recorder utilizes the heat of direct solar radiation to activate the instrument. It comprises a polished solid glass sphere around 10 cm in diameter with the axis mounted parallel to the Earth's north–south axis. The sphere is supported in a spherical bowl so as to bring the image of the Sun to focus on a chemically treated thin card held in a groove inside the bowl. Further, the bowl is an inherent constituent of the spherical shell and surrounds the lower half of the focusing sphere. Importantly enough, the sunshine recorder is not sensitive to weak sunlight. This is the stage or the condition when the Sun happens to be within 5° of the horizon. Simply put, it generally refers to the early mornings and late evenings.

3.6.4.1 Albedometer

This instrument is used to measure both global solar radiation and reflected radiation. Several elements such as mountains, buildings, vegetation's, soil and snow reflect some part of the incident solar radiation. Albedo is a measurement that quantifies how much these mediums reflect the radiation falling on it. This value is usually determined by dividing the total medium reflected solar radiation by the global solar radiation that strikes a designated location.

3.6.5 Measurement Accuracy of Solar Radiation

India has been classified into six different climatic zones on the basis of several key considerations. Climatic conditions that prevail at any given site have impact on the performance of a solar PV system under the actual field operating conditions. Radiation can be measured as follows:

■ The time cycle of radiation measurement is of the order of many years. It simply implies that radiation data are collected over a long period of time.

- Such measurements are generally corrected with the corresponding climatic data of the nearest available meteorological stations in respect of which a long series of observations become available.
- The most probable values of the collected datasets are to be taken for the site under consideration, that is averages, extremes, standard deviations are realized.

3.7 Overview of Solar Radiation Measurement Activity in India

Solar radiation varies throughout the day which makes it all the more important to have an accurate measurement. Such values can be made the best use of in the design, development and performance analysis of the PV power systems such as the solar power plants. Satellite-based irradiation estimates are mostly used for continuous spatial coverage of wide region. Ground-based measurements score easily over those provided by the satellites. The Ministry of New and Renewable Energy (MNRE) has embarked on a massive plan to achieve a target of 20,000 MW of solar PV grid connected power by 2020 under the ambit of Jawaharlal Nehru National Solar Mission (JNNSM).

Table 3.8 shows the state-wise locations of the stations. The Chennai-based National Institute of Wind Energy has initiated an exclusive SRRA unit on a mission mode project and all the 115 SRRA stations stand completed in two phases. To meet the specific challenges in the implementation of JNNSM, MNRE has launched a network of 115 nationwide automatic solar and meteorological measuring stations called the Solar Radiation Resource Assessment (SRRA) Stations (Table 3.8). To implement this project, Centre for Wind Energy Technology (C-WET), Chennai, has started an exclusive SRRA Unit on mission mode project and all 115 SRRA stations completed in two phases. It is also of importance to study the effect of

Table 3.8 Geographical Outreach of Solar Radiation Resource Assessment Centres in India

Synod.	State	Stations (Phase I)	Stations (Phase II)	Total No. of Stations
1.	Andhra Pradesh	03	03	06
2.	Bihar		03	03
3.	Chhattisgarh	01	01	02
4.	Gujarat	11	01	12
5.	Haryana	1	1	2
6.	Himachal Pradesh	2	—	2
7.	Jammu and Kashmir	1	2	3
8.	Jharkhand	3	—	3
9.	Kerala	2	—	2
10.	Madhya Pradesh	3	4	7
11.	Maharashtra	3	6	13
	MEDA SRRA Stations	—	4	
12.	Odisha	4		4
13.	Punjab	2	—	2
14.	Rajasthan	12	1	13
15.	Tamil Nadu	7	1	8
16.	Uttar Pradesh	5	—	5
17.	Uttarakhand	2	—	2
18.	West Bengal	3	—	3
19.	North East Region	10	—	10
20.	Union Territories	1	3	4
21.	Advanced measurement stations		4	4
22.	Total number of stations	51	68	119

Source: Girdhar, J. et al., Solar Radiation Resource Assessment Project in India, June 2014, Akshay Urja Magazine, June 2014, Ministry of New and Renewable Energy [6].

suspended particulate matter (i.e. turbidity/aerosol concentration) in the atmosphere – dust particles, water vapour, gases – on scattering or absorption of solar irradiance. For this specified purpose, Chennai, Kolkata, New Delhi and Gandhinagar are equipped with an advanced measurement station. Each station provides information related to the (a) aerosol column, (b) atmospheric turbidity, (c) column ozone, (d) water vapour and (e) NO_x in the atmosphere.

3.7.1 Salient Features of SRRA

SRRA is a very large-scale project. Each SRRA station comprises two towers with a height of 1.5 and 6 m. The 1.5 m tall tower houses a solar tracker which is equipped with (a) a pyranometer, (b) pyranometer with a shading disc and (c) pyrheliometer to measure the global, diffuse and direct radiation types, respectively. The 6 m tall tower has a wide range of instruments capable of measuring atmospheric parameters of increasing significance for the PV industry. Some parameters are as follows:

■ Ambient temperature
■ Relative humidity
■ Atmospheric pressure
■ Wind speed and direction
■ Rain gauge
■ Data acquisition system

Importantly, all the sensors are traceable to the WMO and World Radiometric Reference (WRR) with high reliability and accuracy. The Sun tracker is configured using a GPS system that always faces the Sun. Solar power is running the SRRA with a 7-day system autonomy feature. On the wavelength monitoring front, there are 10 independent narrow wavelength channels in the range 300–1020 nm.

3.7.1.1 Data Acquisition System

The state-of-the art data acquisition system records 37 measured and derived parameters every second. Data recorded this way is transmitted after averaging them to 1 min directly to the Central Receiving Station (CRS) set up at the C-WET, where fully automated quality control procedure is being adhered for data processing, analysis and report generation. This also includes flagging and gap filling methodology for ensuring quality check algorithms directly applying on the raw data. Importantly, a dedicated level-2 sensor is in place at the National Institute for Wind Energy (NIWE; C-WET previously) to apply algorithms which have been developed for data analysis and quality checks.

As far as the solar radiation value determination like those of Global Horizontal Irradiance (GHI), Direct Normal Irradiance (DNI), and Diffuse Horizontal Irradiance (DHI) is concerned, recourse is taken to application of quality control. Such a control is essentially based on the Baseline Surface Radiation rules as made available by the WMO and elaborated by the Management and Exploitation of Solar Resource Knowledge (MESOR). Further, data on pyrheliometer error (percentage terms), solar elevation and azimuth angles (in degrees), battery voltage and signals on sensor cleaning status are also being received at the CRS, NIWE, Chennai. Additionally, a trigger switch is also installed to track the cleaning status of the SRRA stations on a daily basis. Following the initiation of quality assessment, reports are generated on a daily, monthly and yearly basis. The data gathering and analysis are carried out by a specially designed and implemented software. In CRS, it is possible to monitor data both in numerical and graphical formats.

Importantly, all the parameters from each SRRA station are logged every second, and after averaging for every 10 min, the data are stored in a datalogger. The 1 min data are then transmitted to the CRS set up at NIWE via GPRS communication system. Data thus received at the CRS are processed on a daily basis vis-à-vis all the SRRA stations. Further, the station data are made available to those interested on the NIWE website on a monthly basis. The significant objective of MNRE is to share these value-added data to various stakeholders such as the solar project developers besides academic and research institutions. The server comprising primary and secondary servers is being supplied with uninterrupted power backup for 72 h via two UPS of 8 and 10 kVA. Table 3.8 shows the locations of SRRA centres in India.

3.7.2 Institutional Collaboration

This challenging task of setting up the SRRAs followed up by their data enabling and analysis would not have been possible without the support of a few overseas organizations. Basically, the SRRA project has synergy with the SolMap project, which is funded by the International Climate Initiative (ICI) of the Federal Ministry for the Environment, Nature Conservation and Nuclear Safety (BMU), Government of Germany. SolMap is implemented by GIZ with an active cooperation from the MNRE. Under this arrangement, Deutsche Gesells Chaft for Internationale (GIZ) lends the required technical support and capacity building to the SRRA staff at National Institute of Wind Energy (NIWE). The aim is to achieve and maintain high quality standards vis-à-vis (a) quality measurements, (b) generation of reports and (c) solar atlas. Additional support being offered by GIZ is in terms of scientific analysis of the data coupled with generation of value-added products. The NIWE has also entered into collaboration with the National Renewable Energy Laboratory (U.S. DOE) and Space Applications Centre (Ahmedabad) on matters related to quality checks, solar atlas and, importantly, solar data.

References

1. www.wikipedia.org.
2. Garg, H.P., *Treatise on Solar Energy*, Vol. 1: *Fundamentals of Solar Energy*.
3. Baldocchi, D., Lecture No. 7, Solar radiation, Part 3, Earth–sun geometry, University of California, Berkeley, CA, 2012.
4. Commission of Electrotechnique International/International Electrotechnical Commission.
5. Solar Constant Data Tables consolidated from multiple sources.
6. Girdhar, J. et al., Solar Radiation Resource Assessment Project in India, June 2014. Akshay Urja Magazine, June 2014 Issue Ministry of New and Renewable Energy.

Chapter 4

On-Site–Specific Considerations

4.1 Introduction

Conventional power sources, especially coal-based sources, have led to excessive emissions of greenhouse gas (i.e. carbon dioxide) into the atmosphere, which has led to a serious ecological, social and economic impact across the globe. It is equally true that energy demand has been on a steady rise mainly on account of enhanced population growth coupled with an expanding industry base and geographic distribution. This is where the need of renewable energy (RE) in a far greater measure seems to be inevitable. The increasing utilization of RE is expected to scale up the global energy production to the extent that it would dispense with at least some use of the finite resources. The underlying idea is to protect the environment to the best possible extent. It is in sync with what the clean energy use campaigns championed by several countries across the globe aim at in ultimate terms.

Amongst the RE technologies, solar photovoltaic (PV) energy has attracted a wide attention. This is mainly due to its potential as an energy source with visible benefits for such regions as perceive grid connected power either impractical to extend or costly to use. There has been a focused attention worldwide to expand the production of solar energy within both the developed and developing worlds. Though this energy is of high demand, it contributes only little to the cumulative global energy supply. That is however one side of PV utilization with the other one being the competitive utility scale market, that is the MW power plants connected directly to the locally available grid. Solar PV is slowly but surely inching towards achieving the much needed grid parity and when this happens, solar power capacity would grow manyfold. A noteworthy development in recent years is the availability of most modern solar module manufacturing practices. This has helped to lower down the cost of PV electricity via a multi-pronged approach, that is, via reduced intake of various materials, use of energy-efficient equipment, higher capacity of the processing equipment, reduced energy cost and attaining economies of scale.

As the PV market expands further, manufacturers may continue to standardize the designs and system installations. The significant objective of such innovative considerations is to bring down the cost of PV electricity further and thereby expand the market share to some respectable levels in the future. Today, about 3% of the global energy market is supplied by PV-driven electricity. However, such countries that have prioritized the utilization of RE are able to currently meet nearly 30% of their electricity demand with wind and solar energy resources. Solar to electric conversion efficiencies are moving up as more innovative approaches are being followed up for the grid integration of solar power. All this gives a new hope for solar power taking a quantum leap forward on the global energy front within a foreseeable future.

4.2 Efforts at Preliminary Siting of PV Power Plants

Solar grid-connected power generation avoids the use of expensive and difficult-to-maintain battery energy storage. Today, the focus is not only towards achieving the economies of scale in manufacturing and actual installations but also towards every possible aspect of laying down the PV utility scale power plant. One such key consideration is that of siting the power plant accurately and reliably. Besides observing the spot, the site should be mapped with the best available tools. Table 4.1 lists some site mapping tools available in the market today.

These software tools enable the use of raster (grid based) map data services. This is used mainly to visualize the following information:

■ Solar power potential
■ Critical habitats
■ Development risk
■ Fire potential

The application of geographical information system (GIS) data in the case of RE resource site suitability analysis is a well-established practice being followed globally. It is quite important to select a suitable location for solar power development as it can impact the efficiency of solar power, procurement cost and environmental influences. Factors listed below are generally responsible for the satisfactory siting of solar power projects:

■ Slope
■ Closeness of water and road
■ Ownership of the land area under consideration
■ Grid connectivity

Table 4.1   Site Mapping Tools Available on the Market

Tool	Brief Remarks	Relevant Website Link
Solar Energy Environmental Mapper		www.solarmapper. anl.gov
PV Mapper	Open source GIS-based web application Provides utility scale power developers with tools and data for site selection and screening of potential solar energy plants Used for mapping functions, modelling and analysis	www.pvmapper. org
Landscape Modeler	It is a commercially available web-based tool developed by ESRI Enables a user to specify the environmental and cultural factors deemed to influence the decision-making process Choose suitable data layers, weigh the same as per their importance Utilize the geoprocessing tools so as to identify the best locations for the desired purpose	

In total, solar energy resource analysis for grid-connected power generation projects is influenced by various factors. These could well be classified into four major types such as technical, economic, environmental and social. Importantly, the first three types are dependent on the following few parameters:

- Physical terrain
- Geographic location
- Existing closeness of the available infrastructure to the site
- Land-use restrictions and regulations

Social factor is much in accordance with the changing times and prevalent perceptions and beliefs related to environmental issues. Solar project developers usually frame a site suitability map, which they use to take care of the aforementioned categories. ArGIS software is used for this purpose along with a few more tools available today.

4.2.1 Strategic Approach for Siting

A site found to be abundant in solar energy should be studied for its primary terrain and proximity siting features by experienced professionals. For this, one needs to take into account the prevailing infrastructure which can affect the direct cost of the grid-connected solar power development. It is also important to consider the available solar irradiance as it can influence the efficiency of a functional site – in this case a solar power plant. In a country like India, where the MW scale PV power development has evolved strongly since the last 4–5 years, majority of the plants are located in areas that are far away from the humdrum of big cities and towns. This necessitates a thought on the vehicle access to a developing site. This again is important from the all-important viewpoints/requirements of constructability and associated maintenance.

With road construction being a cost-intensive activity, it is important that a power plant is established at a place where roads already exist. Also the solar power generated at the site has to be transmitted too. This implies that the grid transmission lines should lie close to the site as it can affect the construction and development costs. Further, a large amount of water is required during the development stages of the grid-connected power plants. This is surely in a far less measure once the power plant gets commissioned at a pre-determined site. Only minimum water is required to clean solar PV modules at regular intervals. That of course eliminates the requirement for an average rainy period of 1–2 months per year.

It is pertinent to mention here that a flat terrain is certainly essential for both solar exposure and constructability. Additionally, high solar irradiance is a prerequisite to achieve desired values of both the plant efficiency and stability. Terrain aspect, that is compass direction, can well be regarded as a vital physical parameter for the site selection as it is important in actual site layout and accompanying design issues. Serious efforts to incorporate RE sources are being made across the diverse sectors of the energy economy. This has been necessitated to a large extent because of the following reasons:

■ Increasing demand for electricity worldwide
■ Depletion of conventional generation facilities
■ Consequences of greenhouse gas emissions from existing power plants

One important aspect for achieving such highly ambitious plans is to identify the promising geographical areas. It is equally important to devise a suitable methodology to locate the most potential site(s) within an identified area. The following few parameters are of importance:

■ Availability of solar radiation
■ Availability of unoccupied land
■ Distance from the highways
■ Presence of the transmission lines

Variations in the local climate, module soiling and topography are some criteria that can be excluded. In India,

Gujarat and Rajasthan have received the maximum possible attention so far when it comes to the setting up of large-capacity solar PV grid-connected power generation plants or facilities. Rajasthan could have been chosen because it has the highest available solar radiation in the country. Few studies have been carried out which involve the creation of a series of maps by using GIS software. The underlying objective is to find out the possible locations for establishing large-scale PV power plants. A location chosen in this manner is to be thoroughly analyzed in terms of the following factors:

■ Land morphology
■ Geotechnical status
■ Site preparatory requirements
■ Positioning of the civil structure
■ Creating infrastructural facilities
■ Risk assessment and coverage

In total, interdisciplinary skills and capabilities come under one roof in the development of solar power project. It is an amalgamation of on-site preparedness for putting together a PV plant which is expected to deliver as per the estimated design values of power availability. This chapter takes a close look at the key elements that merit description in an easy-to-understand manner.

4.2.2 Solar Insolation Determination

Within a PV system design, it is important to know the amount of sunlight available at a particular location at a given time. The two most common methods which characterize solar radiation are the solar radiance (or radiation) and solar insolation. Solar radiance is an instantaneous power density expressed in units of kW/m^2. It varies throughout the day from 0 kW/m^2 at night to a maximum of about 1 kW/m^2 and is strongly dependent on the location and locally available weather conditions. Measurements of solar radiance consist of global and/or direct radiation measurements taken periodically throughout the day. The measurements are taken using either a pyrometer (measuring global radiation) and/or a pyrheliometer (measuring direct radiation). In well-established locations, these data have been collected for more than 20 years.

An alternative method of measuring solar radiation, which is not only less accurate but also less expensive, is to use a sunshine recorder. These sunshine recorders (also known as Campbell–Stokes recorders) measure the number of hours in the day during which the sunshine is above a certain level (typically 200 mW/cm^2). Data collected in this way can be used to determine the solar insolation by comparing the measured number of sunshine hours to those based on calculations and including several correction factors. A final method to estimate solar insolation is cloud-cover data taken from the existing satellite images.

While solar irradiance is the most commonly measured parameter, a more common form of radiation data used in system design is the solar insolation. Solar insolation is the total amount of solar energy received at a particular location during a specified time period, often expressed in units of kWh/m^2/day. The units of solar insolation and solar irradiance are both power density (for solar insolation the 'hours' in the numerator are a time measurement as is the 'day' in the denominator). Solar insolation is quite different from the solar irradiance, as the solar insolation is the instantaneous solar irradiance averaged over a given time period. Solar insolation data are commonly used for a simple PV system design, while solar radiance data are used in more complicated PV system performance. In the latter case, the purpose is to calculate the system performance at each point in the day. Solar insolation can also be expressed in units of MJ/m^2 per year.

Solar radiation for a particular location can be given in several ways:

- Typical mean year data for a particular location
- Average daily, monthly or yearly solar insolation for a given location
- Global isoflux contours for a full year, a quarter year or a particular month
- Sunshine hours data
- Solar insolation based on satellite cloud-cover data
- Calculations of solar radiation

In India, most geographical areas have between 250 and 300 sunny days in a year with annual global radiation varying from 1600 to 2200 kWh/m². The Jawaharlal Nehru National Solar Mission (JNNSM) targets a PV power capacity realization of about 20,000 MW by the year 2022. This figure has lately been vastly upgraded to that of 1,00,000 MW by 2022. This places a huge responsibility on the stakeholders, especially the solar power developers, to investigate in suitable land areas to produce quality solar power across the year. A key aspect for preparing ourselves with such a gigantic challenge is to identify the most promising geographical areas for setting up the PV power plants. As the prevailing conversion efficiencies of solar to electric power are low, installation of the solar PV power plant is in need of significant investment in terms of land, capital and importantly, the human resource.

Globally, several works of this nature have suggested methodologies to arrive at the optimal sites vis-à-vis the different types of RE plants. In India, Rajasthan is considered as an ideal area with the following few key attributes:

- About 300–330 sunny days in a year.
- An average daily solar incidence of 5–7 kWh/m²/day.
- A desert area of 2,08,110 km².

- Nearly 60% of the land is arid and semi-arid type.
- Majority of the area is contiguous, relatively flat and undeveloped.

4.3 Solar PV Selection Model in GIS Environment

A GIS is a computer system for capturing, storing and checking accompanied by display of data related to positions on the earth's surface. GIS is capable of showing many different kinds of data on one map. Several models and methodologies have been developed for an optimal selection of sites for the RE technologies including solar PV. One such model enabling the assessment of the suitable sites is the GIS-based model. It is capable of tackling the following elements:

- Topological variation
- Spatial variation
- Weather specific variation

In reality, GIS models operate in a multi-step-process. GIS models are mainly used for the following:

- To identify the criteria that enable the drawing of thematic maps; these criteria impact the site selection process.
- To identify a suitable software support which is well adept at handling the identified criteria.
- To develop software support within the GIS environment.
- To analyze model-specific considerations.

4.3.1 Criteria Identification

The term 'criteria' generally refers to qualifying indicators of a specific process being made actionable. In fact, criteria is to be

identified first and then applied to a desired situation. In case
of solar PV projects, it includes the following:

- Magnitude of solar radiation that strikes the earth's surface
- Availability of unoccupied land (both with regard to its
  existing/future development)
- Accessibility to the site from highways

As is obvious, it is possible to keep the cost of transportation
to the site under control provided the site is in close vicinity of the highway. Importantly, transmission losses can also
be reduced if transmission lines are closer to the site. Taking
a case-specific situation of solar modules into consideration,
these perform well at temperature between 25°C and 45°C.
The following factors also contribute to the ultimate performance of solar modules:

- Possible degradation of cells on account of high ambient
  temperature
- Shadowing of the modules
- Presence of dust on the PV array

It is quite clear that any change in the local climate is deemed
as a vital criterion. Take, for example, the geotechnical issues
which could include due emphasis on the following factors:

- Resistivity of the ground water
- Load-bearing properties
- pH levels of the soil
- Seismic risks

A few more geotechnical issues or considerations are of
importance, especially the geotechnical political issues. This
could involve the nearness of a site to sensitive military zones
and historically significant sites which are better known as
the heritage sites. From a purely topographical point of view,

it is quite preferable to have flat surfaces or slopes that face towards south for projects that are within the northern hemisphere. Another criterion that counts in the existing scheme is the soiling on the module surface. This may reduce the efficiency of a solar PV power plant. It is thus all the more important to take into active consideration the following criteria which could be of a direct relevance:

- Local weather
- Environmental state
- Human intervention
- Proximity of wildlife issues

It is no less important to factor in the miscellaneous criteria such as agricultural activity, dust storms, building activity and traffic

4.3.2 Geographic Data-Specific Software

Currently, several types of software are available in the market to create, manage, analyze and visualize the geographic data. Simply put, geographic data are with specific reference to a location on the earth. It is commonly represented by a term known as GIS software; different types of GIS software mainly include the following:

- Desktop GIS
- Spatial database management system
- Webmap servers
- Server GIS
- Web GIS
- Mobile GIS

However, out of these, desktop GIS software scores easily over the others as it has the capability to create, edit, store, map, intersect and importantly analyze the geographical data.

GIS viewer, GIS editor, GIS analyst are some of its key features. Several companies around the world are quite proficient in making available desktop GIS software, which mainly include Bentley Systems, Erdas Imagine, ESRI, IGIS, besides Intergraph. Several studies suggest that ESRI ArcGIS 10 turns out to be the most suitable choice because of its highly scalable server architecture. The salient features of this widely used software are as follows:

- Offers data in both raster and vector format
- Integrates data from many different sources so as to serve it on its website
- Has a comprehensive system for designing and managing solutions via the specific application of geographic knowledge
- Possesses design capability to automate the many aspects of cartography
- Produces maps quickly

4.3.2.1 Generation of Maps for the Site

Map generation and application of 'Analysis Criteria':
   Cartographic maps can be generated with the help of ArcMap 10.0 using public maps. These inputs which are normally in vector format can be converted into raster format and reclassified. The underlying purpose is to get discrete values from 1 to 9. 1 is the best of all. Thus, mapping is completely normalized.

Cartographic map of vacant land: It can be obtained from the land-use map of a designated location. Any type of land such as barren, waste and vacant land is worth considering. The area required for crystalline silicon technology is around 5 acres for 1 MW PV power plant. So the area required for 10 MW PV power plant should be 50 acres (2,02,342.821 m^2) or more. Thus, cartographic map of a vacant piece of land is an important

Table 4.2   Value Representation for a Solar Road Map

Continuous Value (Distance from Highway in km)	Discretized Value
0–5	1
5–10	2
10–15	3
15–20	4
20–25	5

requirement as it can have several types of diversified features lying within the bare surface.

Cartographic map of proximity from the transmission line: Every state maintains an enriched database of power situation–specific elements. Such a map can be accessed from the concerned authority, say a power distribution company. Take for example, a 10 MW PV power plant which uses a transmission line of 132 kV or more, for which the distance of site from the transmission is considered to be between 10 and 15 km.

Cartographic map of highway: Solar PV power plant sites are generally located away from the highways though not as much as the positioning of standalone systems in the remotest locations of the country. The power map for a particular state is similar to a road map (Table 4.2), which can be studied thoroughly. Accordingly discretized values such as in the following case specific example can be laid down.

4.3.3 Exclusion Criteria

Several maps are now available for ascertaining the best site to obtain optimum levels of PV grid-connected power generation. This typically means intersecting all the maps of analysis criteria and obtaining a class of different locations. By this way, the most suitable region can be scrutinized and settled for vis-à-vis several key considerations. Usually a colour scheme in terms of the following table is put into place to find out the specific location.

Region	Value
Yellow	3
Blue	1
Brown	2

Use of nearby land: It is quite possible that a specific location may have several water bodies such as lakes and streams together with wildlife parks. Solar modules are preferable to be put up in a place where soiling factor is well within the accepted norms. Assume that the region where one wishes to set up a power plant is in close proximity of an urban area or protected area or even a combination of both. Quite clearly, that place would not be an acceptable choice to set up a PV power plant. In such cases, specific colour scheme could be devised to locate the desired place.

Use of a geopolitical site: Solar power plants need to be located with sufficient care. This means choosing sites which are nowhere in the vicinity of international border. Also sites which are in close proximity of the historically significant locations should be avoided. Thus, exclusion criteria of sorts become necessary in this kind of important categorization too. Specific colour scheme can be used in such case to rule out such sites.

4.3.4 Assigned Choice

Solar PV power generation is widely seen as a green power energy source that does not emit any greenhouse gases. For the sake of argument, let us assume a site 'Site A' in a particular state in India emerging as the best possible site to set up a solar PV power plant. Its latitude and longitude coordinates and its climatic data are in accordance with the radiation data sources of NASA, Meteonorm or NREL. The ultimate aim is the electric power generation potential per day – in this case of a 'green region'. This can be deduced by taking into account the following few critical parameters:

- Calculated average annual solar radiation per unit surface per day
- Total suitable area
- Solar to electric conversion efficiency of the solar modules

The following is the mathematical expression to estimate solar power generation potential:

$$(GP, \ kWh) = SR \times CA \times AF \times \eta \qquad (4.1)$$

where
  SR is the annual solar radiation received per unit horizontal area (i.e. kWh/m²/day)
  CA is the calculated gross area of suitable land (i.e. m²)
  AF is the area factor (i.e. it indicates as to what fraction of the calculated areas can be actually covered by the solar modules)
  η is the PV system efficiency

The area factor is chosen to determine the land utilization for the deployment of solar modules without causing any kind of shading. It varies between 70% and 75% of the total available area. Taking into account all the aforementioned parameters, it is quite possible to pinpoint such a significant stretch of land that is highly suitable for solar PV application. Such a region can well be referred to as the 'green region', whose identification also relies on the application of GIS software tool to a good extent.

4.3.5 Simple Concept of a Sun Path Diagram

Sun path diagrams are a convenient way of representing annual changes in the path of the sun through the sky within a single 2D diagram. Solar azimuth and altitude can be read directly for any time of the day and day of the year. They also provide a unique summary of solar position that the designer can refer to when considering shading requirements and design options. The best way to conceptualize a sun path diagram is to liken it to a photograph of the sky, taken whilst lying on your back looking straight up towards the zenith with a 180 degrees fish eye lens. The paths of the sun at different times of the year can then be projected onto this flattened hemisphere.

Each sun path line is generated by determining the exact position of the sun as it passes through the sky in sub-hourly increments for each date – in most cases, on the 1st or 21st of each month. This is then projected from the sky dome onto the flat image.

4.4 Role of Geotechnical Investigation in Solar PV Power Plants

Solar modules are the power producing part of a PV power generation facility. These modules are to be placed on the mounting structures made of hot dip galvanized steel in a series and parallel arrangement. Such structures require grouting in the earth as per the standard norms. For designing various foundation structures of a solar PV project, it is quite an important requirement to undertake a geotechnical investigation. The underlying objective is to study the engineering properties of the soil beneath. This study is commonly known as a geotechnical survey including a topographical survey. Such surveys are better left to the specialized care of the geotechnical consultants. There are several steps involved in the process of a geotechnical examinations and are as follows:

- Sinking a number of bore holes with a diameter of around 150 mm, for example 8–10 bore holes
- Carrying out electrical resistivity tests
- Undertaking standard penetration tests
- Collecting undisturbed tube samples at suitable intervals

The direct outcome of a geotechnical survey is usually represented in terms of the following:

1. Soil profile
2. Bore logs
3. Field test results

It is accompanied by suggesting a suitable type of foundation clearly on the basis of both the laboratory test results as well as the field tests. The field profile is at times altered based on the laboratory test results. This might be due to the poor quality of subsoils, characterized by medium to stiff silty clay/clayey silt followed by very soft to soft clay layer with decomposed matter. Below this, very loose to loose/medium-dense silty sand could be found. This could continue well up to the terminating depth of a few boreholes. Thus, based on the nature of subsoil as revealed from the field tests, suitable type of foundation is recommended. Thus, geotechnical investigation can be divided into the following two parts:

1. Field works
2. Laboratory tests

Field works lead to unfolding of the sub-surface deposit types and their accompanying characteristics, whereas laboratory tests determine the relevant physical and geotechnical properties of the sub-surface deposits. This leads to the firming up of the foundation depths of the structures and the bearing capacity vis-à-vis the sub-surface types and their strength parameters. It is also to be seen in relation to the settlement potentials at the specific site concerned bore holes, etc. Following is a brief account of how the different process steps of geotechnical investigations are normally carried out:

Boring: Boring was carried out by Shell and Auger boring. This generally involves sinking nominal 150 mm diameter bore holes to depths as envisaged by using a mechanical winch. Then the undisturbed soil samples are collected at suitable intervals or at change of strata, whichever is earlier, by open drive sampling method as it was intended to.

Sampling: Nominal undisturbed samples that are 100 mm in diameter are recovered. The sampling equipment used consists of a two-tier assembly of sample tubes 450 mm in length fitted at its lower end. The sampling assembly is driven by means of a jarring link to its full length or as far down as found practicable. The soil is very stiff to hard and contains sand mixtures.

4.4.1 Case-Specific Example

This deals with the soil investigation for a 5 MW PV power plant at NEEPCO, Monarchak site in West Tripura [2]. For any calcareous nodules, cutting shoe was used with an area ratio less than 20%. After withdrawal, the ends of the tubes were sealed with wax and capped before transmission to the laboratory. At close intervals, in-depth disturbed samples were collected for identification and logging purpose. These were tagged and packed in polythene packets and transported to the laboratory.

4.4.1.1 Standard Penetration Test

Standard penetration tests were conducted in the boreholes at intervals of 1.50M/2.0M or at change of strata, whichever is earlier in depth, using a split spoon sampler. The split spoon sampler used is of a standard design having an outer diameter of 50.8 mm and inner diameter of 35 mm, driving with a monkey weighing 63.5 kg, falling freely through 75 cm advances the spoon. A record of the number of blows required to penetrate every 7.5 cm to a maximum depth of 45 cm was made. The first 15 cm of drive is considered to be the seating drive and is neglected. The total blows required for third, fourth, fifth and sixth 7.50 cm of penetration is counted and termed as penetration resistance 'N'. On completion of a test, the split spoon sampler was opened and soil specimens were preserved in polythene bags for logging purpose. All the boreholes were sunk with winch. However, raising of hammer for standard penetration test was done manually. Hence, there will not be any inertia loss, and the efficiency of hammer blows should be considered as 100%.

4.4.1.2 Measurement of Water Table

Standing water level after 24 h of removal of the casing was also noted and shown in the profile.

- Proper site vetting and preparation
- Integrated engineering assessments

4.4.1.3 Summing Up the Geotechnical Issues

Many potential solar projects have not become successful due to feasibility issues. Not surprisingly, identifying a large area suitable for solar plant is the prime requirement for many developers. Without a suitable site, no financial assistance could be availed, even with generous solar incentives.

However, a suitable area is preferred than an open, unobstructed ground space; proper civil, environmental and electrical infrastructure must also exist.

Even relatively simple ground-mounted systems may face multiple challenges. For example, soil composition will determine the size and type of an array's anchoring structure. In other locations, such as reclaimed brownfields or landfills, penetration of any kind may be prohibited. If the area is not relatively flat, it may need to be regarded to accommodate ground-mount racking structures. Distance to the nearest connection point can also be an issue, as long wire runs, especially if, trenched below ground, can tack on significant additional costs. In the current tight margin environment, developers must correctly estimate and optimize system design and site work in order to avoid costly overruns while maintaining a competitive bid.

Most sites require an array of engineering expertise, including civil, structural, electrical and mechanical. In order to ensure optimal system design and site preparation, various project engineers and/or technical consultants must communicate across. Engineering, procurement and construction companies (EPCs) or developers often subcontract different aspects of the project engineering, which can lead to patchwork system design and project delays. Fully integrated EPCs offer a full suite of engineering capabilities and manage utility interconnection study capabilities in-house. This ensures communication between project engineers and product suppliers, whether internal or external, for optimal project execution.

4.5 Overview of Risks in Solar Projects Prior to Their Commissioning

A solar PV project also relies on the construction features like any other project. Thus, construction can be counted as an important risk in case of solar power projects. As per the available information, majority of the solar power projects have experienced low to moderate pre-commissioning risks. On a comparative basis, thermal power projects generally involved higher construction-related complexity. However, solar PV does not involve use of any moving parts. Solar power plants can be safely undertaken even by small to moderate sized companies. The project completion risk is best addressed by having a well laid out EPC contract, which incorporates a fixed price and timeline-specific commitments. It is equally true that a large number of PV project developers choose the well-reliable crystalline silicon-cell technology thus implying a reduced risk assessment.

Another important observation is that long-term power purchase agreements are in place for solar projects thus indicating strong revenue visibility. A noteworthy technical development is the falling price of solar modules over the past few years. The resultant impact can be seen in terms of gradual reduction of solar power tariffs. It is also pertinent to mention here that solar PV sector especially has witnessed an increased market competitiveness mainly due to aggressive bidding by the PV companies. Companies have been relying on the long-term debt periods with tenures ranging between 10 and 13 years. In the time ahead, several factors will exert some positive impact on the credit-risk profiles of PV projects:

- Ability to reduce capital cost
- Ability to reduce the interest rate
- Improve the capacity utilization factor
- Long-term debt availability

Table 4.3 highlights the pre-commissioning and operational risks of solar versus other projects as per the study undertaken by a very prominent credit rating-company known as CRISIL and a leading industry association – PHD Chamber of Commerce.

Table 4.3 Risk Assessment of Solar PV Power Plants

Risk Type	Solar	Wind	Conventional Energy
Project execution risks	Relatively low	Relatively low	Relatively high
Project construction duration	Short (8–12 months)	Short (4–6 months)	Larger time lines (36–48 months)
Resource data risk	Moderate to high	Moderate	Not applicable
Date availability	Operational data for past 2–3 years only	Data for more than 10 years	
Margin of error	Reasonable accuracy obtained	Reasonable accuracy attained	Risk linked to availability of raw material, namely coal and gas
Technology risk	Moderately evolving in nature	Already evolved and tested	Proven and tested
PLF risk	Low variability and seasonality in solar irradiance	High variability and seasonality	Depends on availability and quality of fossil fuels

References

1. Consolidated from multiple sources (including relevant websites).
2. www.bhel.com/dynamic_files.
3. A white paper on India Solar and Wind energy by CRISIL and PHD Chamber of Commerce and Industry, February 2015, www.crisil.com.

Chapter 5

PV System Design Considerations

5.1 Introduction

Solar photovoltaic (PV) systems are generally considered as clean, safe and reliable means to power a given load. Amongst the renewable energy sources such as biomass, wind energy and small hydro power, PV may not always best suit each and every site. However, it has a unique advantage of being the least complex compared to other sources. This is due to the fact that it does not involve any mechanical process unlike other sources. In a solar PV system, mechanical energy is not used in the entire process of power generation. It is quite important though to thoroughly check each potential site and the accompanying system design and engineering requirement. The real question is whether there is a genuine need to give a serious consideration to the system design at all. Well, it makes complete sense to do that keeping in view the following few issues.

High cost per watt – The materials used in device preparation are expensive, and also, processes, such as the development of silicon material, are energy intensive.

Moderate efficiencies – Solar to electric conversion efficiencies can still be termed as being moderate. Only a few cell design configurations made of best chosen tandem material combination such as gallium arsenide and gallium antimonide or for that matter concentrator solar cell designs offer higher efficiencies. However, as such designs are expensive, their use is also limited.

Long-term outdoor stability – Solar modules being the power producing part of a solar system are exposed to outdoor environment throughout the year. Their expected lifespan is between 25 and 30 years and is the non-serviceable part of a system.

Extended payback period – The capacity utilization factor of a solar PV system lies anywhere between 14% and 22%. For example, the cumulative unit generation is 4–5 units/day for a 1 kWp at a site with a peak sun intensity hours of 5.

Clearly, it is of an utmost importance to take into consideration all such factors and design parameters as they ensure maximum possible generation from within a PV system. The underlying objective is to make available a well-performing, safe and reliable system which can run a given load application for its entire lifetime. Now, it is appropriate to deal with various parameters of a solar cell/module which contribute to both the system design and the actual energy generation.

Open circuit voltage (V): Open circuit voltage (V_{oc}) is the maximum voltage that is available from a solar cell. It originates at no value, that is zero value of the current. V_{oc} corresponds to the amount of forward bias on the solar cell junction due to illumination.

Short circuit current: Short circuit current (I_{sc}) is the current that passes via the solar cell when the voltage across the solar cell is zero, in other words, when the solar cell becomes short circuited. It owes its origin to the generation and collection of light-generated carriers. I_{sc} value means a highest value of current as can be drawn from a solar cell.

Maximum power: Maximum power (P_m) is the simple product of maximum values of voltage and current in an I–V curve. Power output of a solar cell increases with voltage, reaching a maximum, and then decreases again.

Fill factor: The fill factor (FF) is defined as the ratio of the maximum power from the actual solar cell to the maximum power from an ideal solar cell. In graphical terms, FF is a measure of the squareness of the solar cell. The more square the curve, the better it is.

Efficiency: Efficiency is defined as the ratio of energy output from the solar cell to the input energy from the sun. It is most commonly used to compare the performance of different types of solar cells. Solar to electric conversion efficiency depends on the solar spectrum, sunlight intensity, and importantly, temperature of the solar cell.

5.2 Key Attributes of Solar Modules

A solar module is the power-producing part of a PV system. In turn, it is composed of a number of solar cells as individual cells are not capable of producing enough power. To remedy this issue, cells and modules are connected together to obtain more power. A solar module basically has a number (e.g. 33 or 36) of interconnected encapsulated cells. It is a weather-proof assembly expected to last for about 25–30 years under the actual field-operating conditions. These modules when connected in series produce a higher voltage, and a parallel connection results in higher current. In general, cells in a solar module possess identical characteristics.

5.2.1 Design and Structure Concept

Solar module was initially conceptualized to charge a normal 12 V lead-acid battery via an interconnection of 36 cells. Commercially available solar cells normally provide an open circuit voltage of about 0.55 V at 25°C. Thus, the cumulative voltage output of 36-series-connected solar cells is 0.55 × 36 = 20 V. Voltage at the maximum power point is equivalent to about 20/1.1 = 18 V. It is appropriate to highlight here the temperature-related effects on solar cells.

- The voltage of a solar cell decreases by about 2.3 mV per degree rise in the temperature of the cell.
- While assuming an ambient temperature of between 30°C and 40°C, the temperature of a solar module under sunlight would be anywhere between 50°C and 80°C.
- As such, the voltage of each cell operating at 700°C would drop by 0.08 V.
- The total drop in the module voltage as a result of high ambient temperature is about 2.9 V, that is, 3 V.

From these explanations, it is quite clear that the operating voltage of a solar module in relation to the aforementioned operating conditions is between 14 and 16 V. It is thus sufficient to charge a 12 V battery.

From a system design perspective, it is appropriate to deal with the wattage of individual solar modules. Solar modules are normally rated in peak watts and produce power from as low as a watt to about 300 Wp or even more. Wattage is a simple multiplying factor of voltage and current. It depends on the current-generating capacity as well as the voltage capacity of a solar module. It is worth mentioning about the pseudo square (mono-crystalline) and truly square shapes (multi-crystalline cells) of a solar cells. The packing density of a solar module is defined as the percentage of the cell area in the entire module area. It basically depends on the shape of the solar cells. Circular cells offer a packaging density of 70% as against 80% for pseudo-square shaped cells. Square solar cells enable the highest possible packaging density of 90%. The effect of packaging density on the power output of a solar module is evident.

Interestingly, current from a module is directly proportional to its size, but voltage is not dependent on module size. The importance of current and voltage in case of a solar module is illustrated in the following:

- The current-generating capacity of a silicon solar cell is about 30 mA/cm^2 at 1000 W/m^2. Solar cells are commercially available in two sizes – 12.5 × 12.5 cm^2 and 15 × 15 cm^2.
- Thus, a silicon solar cell of size 15 × 15 cm^2 will produce a current of about 6.75 A.
- Taking the voltage and current values into active consideration, a module comprising 36 solar cells is expected to deliver a peak power rating of 15 × 6.75 = 100 Wp.

5.2.2 Industry Ratings of Solar Modules

Solar modules exhibit their performance in direct relation to the outdoor environment of a given site. Modules are usually rated in terms of their peak power (Wp) output under standard test conditions (STC). In turn, STC refers to solar irradiation of 1000 W/m^2, AM 1.5 Global solar radiation together with cell or module temperature of 25°C and wind speed equivalent to 1 m/s. As outdoor STC may not always be met, nominal operating conditions (NOC) are taken into consideration according to which irradiation value is equivalent to 800 W/m^2, ambient temperature is 20°C together with a wind speed of 1 m/s.

In this case, nominal operating cell temperature (NOCT) is considered. It is defined as the temperature attained by a cell in open circuited module under NOC conditions. Furthermore, this is used to give more accurate cell temperature of the module under operating conditions. The normally observed value of NOCT lies anywhere between 42°C and 50°C. Mathematical expression used for the specified purpose is as follows:

$$T_{mod} = T_{amb} + (NOCT - 20/0.8) * P_{in},$$

where

T_{amb} is the ambient temperature
P_{in} is the solar irradiation in kW/m²

5.2.2.1 Effect of Solar Irradiation

Sunshine available during the day is the prime driver of power output availability from a solar module. Solar irradiation has a direct influence on the current generated by a solar module. Current is a linear function of the radiation intensity. The power of solar module decreases almost linearly with a decrease in the intensity of solar radiation. It is a fact that the solar irradiation available throughout the day is changing. In contrast, voltage of a solar module is a logarithmic function of the radiation intensity. Therefore, it is evident that power output changes with a change in radiation.

5.2.2.2 Effect of Temperature

Solar modules are placed outdoor throughout the day. The output power of the solar power modules also depends on the temperature at which the module is running. The increased cell temperature decreases the open circuit voltage, which is basically due to an increase in reverse saturation current. Further, the peak power decreases with an increase in module temperature.

5.2.2.3 Mismatch in Solar Cell/Module

Though solar cells and modules are currently being manufactured in modern production set-ups, all the cells that are produced are not similar in terms of electrical parameters. The following are a few underlying reasons:

■ Variation in the processing of solar cells.
■ Cells of an equivalent power rating of different makes. That means two different sources of manufacture.
■ External conditions experienced by the cells/modules may be different which may cause their shading.
■ Exposure to ultraviolet light (UV) may lead to damage. This makes the encapsulating material of the cell to become semi-transparent from its original stage of being transparent.

5.2.2.4 Mismatch of Series-Connected Solar Cells

Power loss due to mismatch depends on the operating point of a solar module and the degree of mismatch. Additionally current of the combination is the short circuit current of the poor cell. Thus, surplus power produced by a well-performing cell is simply dissipated into a poor cell, which leads to irreversible damage. Compare this with the resultant effects in an open-circuit voltage scenario. Significant power loss could take place at low voltages. This mismatch is primarily due to differences in the open circuit voltage and differences in the short circuit current.

5.2.2.5 Occurrence of Hotspots in Solar Modules

Within a solar module are a number of series-connected solar cells. Under short circuit condition of the string, the shaded cell becomes reverse biased. A hotspot condition occurs when there is one low current cell in a string of at least several high short circuit current solar cells. One shaded cell in a string reduces the current that flows through the good cells. This makes a good cell to generate higher voltage with a risk of often reverse bias of a bad cell.

5.2.2.5.1 Use of Bypass Diode

A bypass diode or two is generally added in a solar module, whose key function is to bypass the current. A partial shading brings down the forward bias voltage. In case of shading, current from a string of cells is limited by the cell that carries the lowest current. If some cells get shaded, then the surplus current from the good cells in the string forward biases these cells. It is a general practice to use two bypass diodes for a module comprising 36 cells.

5.2.2.5.2 Use of Blocking Diodes

In dark conditions when no voltage is being produced by the solar modules, the voltage of the battery would cause a current to flow in the opposite direction through the modules, thus discharging the battery. Blocking diodes are usually used to prevent this type of situation.

5.3 Current–Voltage Characteristics

An I–V curve of each module for any end-use application is available in the market. Following are a few important observations to be considered:

- Power output increases with an increase in module voltage.
- Power output reaches a peak and is commonly known as the P_{max}.
- Power output drops as the voltage nears the open circuit voltage value.

Various factors exert their influence in a varying measure when it comes to assessing the suitability of a site for PV system installation. The main factors are discussed next.

5.3.1 *Availability of Solar Energy*

Sunshine is generally not present in an equal measure on all the places on Earth. Therefore, one has to know whether there is enough sunlight available at the site of choice and if so whether is it available for most parts of the year, for example, around 75% of the time period. There is a likelihood of having a site that is shaded by trees or that is in the shade of a mountain; a site may also be surrounded by buildings which may cause shade during the actual working of a PV system. If so, there is a need to look for a different site.

5.3.2 *Cost of the System*

A solar PV system is quite easy to maintain during its lifetime. The power-producing part, that is the solar module, should be cleaned at regular intervals. It is akin to dusting off your other appliances such as an LED TV or a laptop at home. PV systems cost little to operate on a daily basis but do need routine checks and care. Solar PV systems even today have a relatively high initial capital cost, although there has been a continuous downward cost trend more so during the past few years. At the same time, the central government and the state governments offer financial and fiscal incentives to make these systems financially attractive. It is also true that PV system is surely less expensive than running a diesel generator for instance. Solar systems are mostly opted in rural areas due to their sheer need. This is because the cost of extending grids to such areas having small loads and dispersed load demand is usually prohibitive. In India, for example, it is now an attractive proposition to set up grid-connected PV power plants in an MW capacity range. Today, it costs only 55 million rupees to set up a 1 MW PV power plant compared to 180 million rupees some years back. Likewise, it costs around Rs. 75,000 to install a PV-grid type power plant of capacity 1 kWp, whereas to install with a battery storage, the cost incurred is about 25% more.

5.4 Additional Attributes of a Solar PV System

Modularity: Majority of the PV systems can be expanded so as to take care of enhanced loads. However, once the PV array is installed, it hardly undergoes any change till its expected lifespan. In some cases, however, new modules are added in order to offset the impact caused by the routine annual degradation of the solar modules.

Reliability: Solar PV systems ensure reliable operation owing to efficiency improvements at the key component levels. Design and installation practices have bridged the gap between the simulated results and the actual realization of the parameters of significance.

Flexibility of approach: PV modules had a high initial capital cost which necessitated the need for hybrid combinations of PV with other available sources of renewable energy. These types of systems are better known as the hybrid systems. Take, for example, the combination of PV–wind energy as a preferred application accompanied by a few more hybrid combinations such as PV–biomass hybrid system, PV–wind energy or PV–hydro. Such systems give a good opportunity to extract the best attributes of each energy resource.

Before discussing the system design procedures without a recourse to the simulation techniques, it is especially relevant to discuss a few more parameters.

5.5 System Design–Specific Considerations

Site worthiness: It is quite an important requirement to figure out whether a potential site is a suitable choice for a PV system. This should be done based on both the present and future scenarios. The relevant factors vis-à-vis ascertaining the site suitability are as follows.

Solar radiation exposure: Solar modules are the power-producing part of a solar PV system. Therefore, it is important for them to receive a maximum possible amount of direct sunshine throughout the day. Major portion of the available energy is obtained from the sun normally between 8.30 and 3.30 p.m.

Obstructive elements: Solar modules are best positioned when they face towards the south. Thus, it is important to ensure that the southern exposure of these modules should be as unobstructed as possible. Obstructions could include, for instance, buildings, trees, and mountains besides other objects that may shade the PV array modules. Unused power lines may pose some obstructive problems.

Terrain and space considerations: Solar modules can be placed either on the ground or on the available roof space. There is one more choice which, though, does not find favour with a majority of installations worldwide. Such a type of application is commonly known as the building integrated photovoltaics or simply BIPV. In this type of design configuration, solar modules are architecturally integrated into the building envelope. Terrains whether flat or steep are an important consideration as flat surfaces normally produce feasible amounts. Rocky sites such as mountainsides are normally ruled out but are used in unavoidable situations. Enough space should be left to place the PV arrays safely and properly to avoid any type of long-term maintenance and to ensure safety.

Weather considerations: Bright sunshine is ideal for a good field performance of a PV system. However, a suddenly appearing cloud patch could dramatically reduce the solar insolation at any given point of time and therefore PV modules may produce lesser amount of energy on a rainy day. Temperature happens to be a pivotal consideration and thus an important design cum performance issue. It is a well-known fact that solar cells perform the best in cooler and clear conditions. Higher temperatures often result in a reduced power output from the PV array, thereby lowering the system efficiency as a whole. Proper attention needs to be accorded to module selection, string sizing and balance of system design considerations.

5.6 Balance of System

A set of components of the PV system other than a solar module is generally known as the balance of system (BOS). The BOS normally includes module-mounting structures, batteries, charge controller, inverter, wires (i.e. the conductor), conduit, earthing circuit, fuse disconnects and so on. It must be chosen with sufficient care for each such system as enables one end-use application or the other. Further the BOS should be properly sized so as to ensure smooth running of the complete system and keep the losses to the bare minimum.

5.6.1 Matching Issues with System Components

A solar PV system is capable of converting incident sunlight into useful electricity. Key system components include solar modules and power-conditioning unit. Let us assume the modelled energy to be X with Y as energy delivered to the load. Then the system performance ratio is equal to Y/X. In general, performance ratio of a system depends on the following parameters:

- Solar modules
- Power-conditioning components and their interface/ integration
- Environment

Batteries store the energy generated during the day and run the designated load via an efficient charge controller.

Following are a few design configurations that are available on the market.

Shunt controller: Solar PV runs at the maximum power point where the surplus power is utilized to run the auxiliary load.

Series controller: A series controller disables further current flow into batteries when they are full. In this case, solar PV runs with the power which is needed to charge a battery. Following which, the battery runs the load via a controller.

Solar inverter charger: In this case, PV runs at a maximum power point so as to charge battery and supply load.

Pulse width modulation controller: A pulse width modulation controller maintains the voltage required for charging a battery.

Maximum power point controller: A maximum power point controller derives the maximum possible PV power.

5.6.2 Key Design Characteristics

Charge controller: Charge controller is a vital interface between a solar module and battery. Therefore, this unit should be chosen with sufficient care. The modern-day charge controllers are designed to meet several characteristics features and these include the following:

■ The battery should be protected against overcharging and complete battery drain to maximize its battery life.
■ The controller voltage must match the voltage available from a battery.
■ The controller input current must be 125% more than the array short circuit current.
■ Programmable charge profile should suit the battery characteristics.
■ A battery temperature sensor along with charge voltage compensation should be present.

- A provision for automatic battery voltage selection (say 12/24 V) should be available.
- Other desirable features such as low idle current consumption and high power conversion efficiency should be enabled.
- On the safety/protection side, a high-voltage disconnect, low-voltage disconnect and, importantly, short circuit/overload protection should be present.

These characteristics must be met during the design process while choosing the rating, make and operational capability of the above nature.

Pulse Width Modulation Controller

- It is important to have a perfect match between nominal PV voltage and battery voltage.
- Power conversion efficiency ranges between 70% and 80%.
- Usual capacity range progresses up to 48 V/60 A.
- It is not a critical need to have well-matched solar modules in a string.
- Pulse width modulation controller is of a small size, not modular and affordable.

Maximum Power Point Controller: The primary concept of maximum power point tracking (MPPT) hinges around an automatic way of finding PV operating points to obtain a maximum DC power output (i.e. P_{mp}). The maximum power point changes as a result of the changing values of irradiance and module temperature and has the following key characteristics:

- MPPT electronic unit allows the PV voltage to exceed the battery voltage.
- A higher voltage/current rating is possible in addition to a higher power conversion efficiency ranging between 95% and 98%.

- Availability of higher voltage/current ratings.
- Critical requirement of matching of modules in a string.
- These types of controllers appear in a bigger size and can be scaled up.
- The cost of a MPPT controller is two to three times more than that of a pulse width modulation (PWM) controller.

5.6.3 Underlying Attributes of an MPPT Controller Technology

Solar radiation keeps changing across the year. Low irradiance is observed especially during the monsoon time along with changing temperature conditions. Additional power becomes available as enhanced battery charge current or grid feed current. Furthermore, constant power output under stable operating conditions leads to an increase in charge current for lower battery voltage. Basically, the MPPT feature leads to an earlier activation of the solar water pumping, thus implying enhanced pumping hours. As per the observational data, MPPT can raise the output by 10%–15% under NOCT conditions and 20%–25% at low irradiance.

5.7 Sizing of PV System

Sizing of a PV system means determining as to how much energy is required to run the system and how many PV modules are actually needed to generate it. A PV system has to generate enough energy so as to cover the energy consumption of the loads (lights, appliances, etc.). The PV system is usually designed by using a computer model. First the energy yield of the PV system for a particular location is calculated. The size and configuration of solar array are then optimized in order to match the energy yield of the system to the energy consumption of the system. The energy yield of a PV system depends on the following parametric elements:

■ Meteorological conditions at the site
■ Type of solar modules
■ Orientation of the modules
■ Characteristics of a PV inverter

5.7.1 Manual Procedures for System Sizing

Simulated models, that is, spreadsheet-based computer models, may not be always available. Therefore, the size of a PV system can be calculated by using an available set of formulae and reference range(s) for various parameters. Site-specific variations are present which need to be included in the process of manual system designing. If such models are not available, then a rough estimate of the sizing of a PV array and batteries can be made by using the following design rules:

■ Determine the total load current and operational time
■ Add system losses
■ Determine the solar irradiation in daily equivalent sunny hours
■ Evaluate the total solar array current requirements
■ Work out an optimum module arrangement for solar array
■ Determine battery size for recommended reserve time, that is autonomy of the system which simply means taking care of the no-sunshine days at a specified site

5.7.1.1 Determine the Total Load Current and Operational Time

Prior to determining the current requirements of loads of a PV system, one has to decide the nominal operational voltage of a PV system. Usually, one can choose a nominal voltage between 12 and 24 V. The next step is to express the daily energy requirements of loads in terms of current and average operational time expressed in Ampere-hours (Ah). In case of DC loads, the daily energy (Wh) requirement is calculated by multiplying the power rating (W) of an individual appliance with the average daily operational time (h). Finally, by dividing Wh by the nominal PV system operational voltage, the required Ah of the appliance is obtained.

Case-Specific Example – I

A 12 V PV system consists of two DC appliances – A1 and A2. They have a power requirement of 15 and 20 W, respectively. The average running time on a daily basis is 6 h for device A1 and 3 h for device A2. The daily energy requirements of the devices expressed in Ah are calculated as follows:

Device A1: 15 W × 6 h = 90 Wh

Device A2: 20 W × 3 h = 60 Wh

Total: 90 Wh + 60 Wh = 150 Wh, 150 Wh/12 V = *12.5* Ah

In case of AC loads, energy use has to be expressed first similar to the DC energy requirement. This is because PV modules generate DC electricity. The DC equivalent of the energy use of an AC load is determined by dividing the AC load energy use by the efficiency of an inverter, which is typically 85% for an off-grid PV use. By dividing the DC energy requirement by the nominal PV system voltage, the Ah is determined.

Case-Specific Example – II

The line of treatment differs in terms of different appliances use, for example computer and television in this case. Let us consider an AC computer (A3) and TV set (A4) connected to a PV system. The computer, which has rated power as 40 W, runs 2 h/day, and the TV set with rated power 60 W runs

3 h/day. The daily energy requirements of these two appliances as expressed in DC Ah are calculated as follows:

Appliance A3: 40 W × 2 h = 80 Wh

Appliance A4: 60 W × 3 h = 180 Wh

Total: 80 Wh + 180 Wh = 260 Wh

DC requirement: 260 Wh/0.85 = 306 Wh

In order to get this figure in Ah, the Wh value is divided by the nominal system voltage, which is 12 V in this case.

306 Wh/12 V = *25.5* Ah

5.7.1.2 Add System Losses

Some components of the PV system, such as charge regulators and batteries, use energy to perform their functions. We denote the use of energy by the system components as system energy losses. Therefore, the total energy requirements of loads, which were determined in step 1, are increased by a factor of 20%–30% in order to compensate for the system losses.

Case-Specific Example

The total DC requirements of loads plus the system losses (20%) are determined as follows:

(12.5 Ah + 25.5 Ah) × 1.2 = 45.6 Ah

5.7.1.3 Determine the Solar Irradiation in Daily Equivalent Sun Hours

The Sun does not shine in an equally bright manner throughout the day. In fact, the solar intensity varies rising from almost nil at sunrise to its peak value at noon. After noon, it starts to diminish until sunset. Thus, the amount of energy that a solar module actually delivers depends on several factors. These mainly include the following:

1. Soiling factor, that is the presence of dust/dirt at the site
2. Local weather
3. Component-level efficiencies

No less important is the proper installation of solar modules. Solar energy collection is bound to suffer if the module orientation/tilt is not chosen accurately. Thus, a solar module should be installed at the correct tilt angle in order to achieve a best year-round field performance. The energy produced in winter is generally much less than the yearly average, whereas the energy produced in summer is more than that of the yearly average. One equivalent Sun means a solar irradiance of 1000 W/m^2. This value relates to the standard at which the performance of both the solar cells and modules is tested under a sun simulator. The daily average solar radiation incident over India varies between 4–7 kWh/m^2.

5.7.1.4 Case-Specific Example

Let us consider a location where the peak sun hours are around 5 per day. There is a 1 kWp PV power pack installed at that location. Thus, it simply means an annual production of around $5 \times 300 = 1500$ kWh, where the number of sunny days is 300 on a yearly basis. Any change in this value affects the net energy generation from a given system.

5.7.2 Determine Total Solar Array Current Requirements

Solar array current determination is quite critical to system design and engineering. The current that has to be produced by a given solar PV array is normally found by dividing the total DC energy requirement of the PV system including the loads and system losses (as calculated in Section 5.7.1.2 and expressed in Ah) by the daily equivalent sun hours.

Case-Specific Example

In the load profile example discussed, the total DC requirement of load including the system losses was 45.6 Ah. Assuming a location with peak sun hours of around 4 h, the required total current generated by the solar array is calculated to be 45.6/4 = 11.4 A.

5.7.2.1 Determine Optimum Module Arrangement for Solar Array

It is generally observed that the PV module manufacturers produce solar modules which may or may not have exactly the same amount of output power characteristics. Thus, an optimum arrangement of modules is the one that will produce the total solar current with the minimum number of modules. Modules are normally connected in a series and parallel arrangement. On connecting the modules in series, nominal voltage is increased, whereas when they are connected in parallel, current gets enhanced. The series and parallel module numbers are calculated as follows:

1. The number of modules in parallel is calculated by dividing the total current required from the solar array by the current generated by the module at peak power (this value is mentioned in the specification sheet).
2. The number of modules in series is determined by dividing the nominal PV system voltage with the nominal module voltage (this value is mentioned in the specifications sheet).
3. The total number of modules is the multiplying factor of number of series- and parallel-connected modules.

5.7.3 Case-Specific Example

Total current produced by the solar array is equal to 15.2 A.
Take, for example, a solar module whose current value as per
the manufacturer's specification sheet is 3.15 A. The number
of modules as determined is 15.2/3.15 A = 4.8. Nominal voltage of the PV system is 12 V with the nominal module voltage
being 12 V. Thus, the number of solar modules is simply put
as 12/12 = 1 module. Hence, the total number of modules that
would constitute a PV array is equal to 5 × 1 = 5.

5.7.3.1 Determine Battery Size

Batteries are generally regarded both as a vital/weak link of
a PV system. These form a key component of a stand-alone
PV system. The primary role of the batteries is to enable
load operation during the non-sunshine hours. For a reliable
operation of a PV system, it is a must to take into account the
cloudy days as well. Therefore, reserve energy capacity should
be built within the batteries. The suitable term coined for this
specific purpose is 'system autonomy.'

The number of reserve days that is generally considered
ranges between 1 and 3 much in accordance with the criticality
of load operation. Simply defined, autonomy is a period of
time during which the PV system is not dependent on energy
produced by the solar modules. It is commonly rated in days.
Take for example the PV system need in telecommunications
or health sector. In both these cases, a reserve capacity of at
least 3 days seems to be preferable. The battery capacity is
determined by multiplying the daily total DC energy requirement of the PV system including loads and system losses (as
calculated in step 2 and expressed in Ah) by the number of
days of recommended reserve time. As a recommended measure, batteries should use just about 80% of its rated capacity.
Thus, the minimal capacity of the batteries is determined by
dividing the required capacity by a factor of 0.8.

Case-Specific Example

Cumulative DC requirement of load including system losses is equal to 45.6 Ah. Recommended reserve time capacity for a specific installation under consideration is around 4 days. Thus, the capacity of a battery is equal to 45.6 × 4 = 182.4 Ah. Accordingly, the minimal battery capacity for a safe and reliable operation is 182.4 Ah/0.8 = 228 Ah. Simply put, the battery takes care of the load for a total duration of 4 days when no sunshine is available.

5.8 Cost of a PV System

Solar modules accounted for as much as 55% of the total cost of a PV system until a few years back. The scenario has now changed due to the fast decline in the cost of solar modules. Therefore, the cost of modules in an off-grid solar system has come down to as low as 35%–40%. In general, cost of a solar PV system depends on the cost of the following few components:

- Solar modules/array
- Balance of the system
- Transportation and insurance
- Applicable taxes and duties
- Installation and commissioning costs

The PV system cost also includes the cost of project management and system design and engineering. The relative contributions of these costs to the total price of an installed system depend on the application, the size of the system and, importantly, the location.

Cost contribution of solar modules: PV applications are currently being promoted under various schemes both for the off-grid and online grid uses. In case of a grid-connected installation, modules turn out to be the major cost contributor.

The cost of solar modules can be reduced based on the following:

1. Enhanced solar to electric conversion efficiencies of solar modules
2. Improved manufacturing techniques (leading to higher system throughput, lower material losses, better yield, etc.)
3. Economies of scale in production

It is foreseen that the cost of solar modules will continue to fall for some more years to come. Retail price of the modules also depends on the number of modules being supplied. Modules can become cheaper if they are proved to be a replaceable element. Take, for example, building-integrated PV. Special modules are now available which can be architecturally integrated into the building envelope. These modules could well be developed for integration into the facades of the buildings, with which it is possible to offset the costs against the costs of the cladding materials, for example, as a replaceable element. Likewise, modules are being utilised to act as sound barriers for the motorways, etc.

5.8.1 Cost of the Balance of System

The cost of energy storage made available in terms of battery use is maintaining an upward trend mainly because of the costs of materials such as lead, zinc and calcium. Because of this the batteries for a stand-alone PV system turn out to be costlier. The cost of a battery depends to a significant extent on the type and quality. The expected lifespan of a majority of solar batteries ranges between 4 and 7 years, which implies that system designers must calculate their replacement cost during the entire lifetime of a solar system. In the case of a grid-connected PV system, the key BOS cost is that of inverter and grid interface unit.

These costs too have come down substantially in India. At the onset of the JNNSM, the peak watt cost of an inverter was around Rs. 55 and this has now come to as low as Rs. 7 per Wp. Several reasons contribute to this dramatic decline of the inverter cost. Increased competitiveness amongst the foreign companies together with an expanding market volume helped to realize such low costs. In addition, few major inverter manufacturers have set up their inverter production units in India, which also contributed to the overall cost reduction. Inverter units also come with an expected lifespan of above 10–12 years, which means they need to be replaced just once during the lifetime of a PV plant.

Chapter 6

Introduction to PV System Sizing Procedures via Simulation Software

6.1 Introduction

Simulation modelling and analysis have become increasingly popular as a technique for improving or investigating process performance. Simulation has found a wide range of applications in various sectors, for example healthcare, computer and communication system, manufacturing and material handling system, automobile industry in addition to logistics and transportation system, service system, military and scheduling. Since simulation is a main tool for modelling and analysis, technology advancement in this field is rapidly increasing. Recent advances in simulation methodologies, availability of software and technical developments have made simulation one of the most widely used and accepted tools in system analysis and operation research. An increased number of simulation software products are available today on the market. Simulation is being widely used now because the software have become cheap and more user-friendly. That is the reason why task-specific simulation software are available in the market. In the field of solar PV system, design of a stand-alone PV system requires customized simulation software based solution as per the requirement of the users.

6.2 Trends on Advanced Simulation Software Development for PV Installation

Designing a solar PV system is based on (a) the best drawn practices in site resource assessment, (b) quality component selection and (c) optimum component engineering and well-designed component assembly from all possible consider-ations. The earlier trend of manually calculating the system size is fast being replaced by the new-generation energy-simulation techniques. The application of thumb rules to work out the quantum of energy generation is no longer relevant except for a speedy approximation of the system size. Simulation in simple words means playing around with differ-ent sets of values that are relevant to the system design under active consideration. The following section gives a bird's eye view on the available range of both commercially and non-commercially used simulation procedures.

6.2.1 Types of Simulation Software

There are 12 major types of software for simulating solar PV systems:

- RETScreen
- PV F-Chart
- SolarDesign Tool
- INSEL
- TRNSYS
- NREL Solar Advisor Model

- ESP-r 11.5
- PVSYST 4.33
- SolarPro
- PV-DesignPro-G
- PV*SOL Expert
- HOMER

Other simulation software available today include Polysun, APOS Photovoltaic StatLab, PV Designer, Solar Nexus, Valentin Software, PV Cost simulation tool, NREL's In My Backyard, Solmetric Suneye and Blue Oak Energy.

6.2.2 Comparative Features of the Simulation Software

The main features of the popular PV simulation software are highlighted here. These software are basically commercial tools which are dedicated to the design of PV systems connected to the grid or in a remote area. An accurate evaluation of the energy output of solar PV panels is obtained by these software. Most of these softwares are available as trial versions for up to a month. The following section explains briefly the salient features of the PV simulation software.

6.2.2.1 RETScreen

RETScreen is a decision-support software developed in Canada which provides the facility of a complete database for virtually any location across the world. This is specially optimized for use with the best available data at each location from about 20 sources. Out of these, the most important ones are those of World Radiation Data Centre (WRDC) and NASA Irradiance data. Further, temperatures and wind velocities are also provided most likely with good reliability. Both these data sources are available at no charge. The significant objective of the software is to estimate the following parameters:

- Estimation of energy production and savings
- Cost assessment
- Emission reductions
- Financial viability
- Risk assessment for different types of renewable energy and energy-efficient technologies

In total, it enables modelling and analyzing of any clean energy projects for a wide range of stakeholders. These may include architects, engineers and financial planners.

6.2.2.2 Easy Solar

This enables a user to design and prepare customized and professional offers within a short duration. This has become possible due to cloud technology which can be interfaced with smart phones to manage sales figures. In fact, the user-friendly application creates better communication amongst the working team with its integrated chat and workflow tool. Thus, it is possible to develop flexible designs on images, Google maps or even sketches. This software offers an additional facility of having in-built measurement tools which allow the designers verify azimuth and inclination of the roof for specific locations.

6.2.2.3 PV F-Chart

This specific software can be termed as a comprehensive PV system analysis and design program. This provides monthly average performance estimates corresponding to each hour of the day. Further, the calculations are based upon the methods developed at the University of Wisconsin. These make use of solar radiation usability to account for the statistical variations of the radiation and the load. The techniques used in calculation reveal statistical variation of radiation and the load taking into account the sheer usefulness of solar radiation. Following are a few key attributes of this software:

- Weather data of more than 300 locations with a capacity to accommodate additional weather data
- Model utility interface systems, battery storage systems and stand-alone systems
- Quick execution of hourly load profiles for each month, statistical load variation and life-cycle economics with cash flow

6.2.2.4 Plan4Solar PV

This provides an all-in-one solution for simple and quick planning and calculation of PV systems. It also involves a comprehensive assembly and connection plan besides all relevant calculations such as inverter design, revenue forecast and profitability calculations.

6.2.2.5 BlueSol

This software is used across the globe to design solar PV systems. This enables a user to carry out the complete process of designing a PV system, that is ranging from the initial assessment of producibility to the actual realization of the project documentation. This software has been made available with a standard Microsoft interface and easy-to-use features.

6.2.2.6 PV Scout 2.0 Premium

This is a PV system sizing software mainly used for the planning and calculation of grid-connected PV systems regardless of the make. In simple terms, it provides a graphical representation and dynamic interconnection of several roofs.

This gives a complete range of data regardless of the product make. Further, the solar radiation data utilized are based on high-resolution mean-minute values. Calculations of economic efficiency are included in the system. In total, the software enables the availability of technical details, rate quotation, electric circuit diagram, DC connection drawing apart from the all-important values of solar yield and economic efficiency.

6.2.2.7 Transient Systems Simulation

This specific software with wide applicability both in the solar thermal and in solar PV application areas took shape way back in 1975. It is a collaborative product of countries such as the United States, France and Germany. Transient Systems Simulation (TRNSYS) alleviates the accession of mathematical models together with the following features:

- Potentialities of the multi-zone building model
- Usable add-on components
- Ability to port with other simulation programs

The specific end user chooses a system description language so as to specify the components for system build-up. This software finds immense use across several segments.

6.2.2.8 PV*SOL Expert

This software is used to visualize many types of roof-parallel and roof-integrated systems. The underlying purpose is also to determine the shading on the basis of 3D objects. The number of modules placed by it in 3D runs up to around 2000. The software makes a reliable computation of the average frequency of the modules shadowed by the objects in addition to a graphical representation of the results derived. It is also possible to find out the shadows which are cast at different times of the day and year. These have some influence on reducing the average yield available from the design facility. Significant objectives of the software are as follows:

■ Visualization of annual irradiation decrease for each point of the PV area
■ Optimization of the solar module coverage and configuration much in accordance with the shading position
■ Automatic and manual solar module roof coverage
■ Output of the simulation of shading in small intervals of 10 min each
■ Yield-based simulation based on the accurate shading ratio determined for each module

Overall, this software is a sound tool to obtain animated visualization of the course of shade at any point in time. This makes it unique from the commonly used simulation software.

6.2.2.9 PVSYST

This has an enviable distinction of being one of the oldest PV software. It has been developed by the University of Geneva and is designed to be used by the architects, engineers and researchers alike. PVSYST offers a wide range of features which increase the overall usefulness and outreach of this specific software. The following are some of its features:

■ Presents three different choices such as preliminary design, project design and tools
■ Complete design availability of the remote PV systems
■ Complete design of PV systems connected to the grid
■ Complete database of solar modules and inverters
■ Provides meteorological data for a large number of stations across the world
■ Useful 3D application to simulated near shadings

- Import of irradiation data from PVGIS and NASA databases besides import of solar module data from Photon International
- Enables economic evaluation and payback in addition to export of calculations to a comma separated values (CSV) files

Apart from these, several other tools are available within this software to simulate the behaviour of modules and cells much in accordance mainly with the irradiation, temperature and shadings. It also contains an expanded contextual help which explains clearly the procedures and models used. This software is able to import meteo data from several varied sources as well as personal data. The final outcomes, that is the results, are available in the form of a complete report, specific graphs and tables. This software has a trial period of 1 month, during which period the full version is accessible. The data of certain stations are included and new data sets can be created by importing data. Further, the software has a preliminary and a project design mode and the preliminary mode can be used to get an approximate value of radiation and power output of system. The project design mode also allows user-defined losses and inverter efficiency in addition to shading analysis. There are many other variables which provide a more accurate output. Three main modules that are incorporated in the software are the preliminary design, project design and tools.

This software is well suited for stand-alone, DC-grid systems and grid-connected and water pumping systems.

6.2.3 Other Simulation Software

6.2.3.1 Solar Advisor Model

National Renewable Energy Laboratory (NREL) is a prestigious laboratory of the Department of Energy, United States. NREL has developed a software named Solar Advisory Model (SAM) as a performance and economic model. It operates on the TRNSYS engine. The underlying purpose of designing this specific model is to enable decision making for the stakeholders of renewable energy industry/programme. It is desirable to make use of this software in tandem with technology and cost benchmarking besides penetration analysis. In essence, SAM puts to use a systems-driven approach (SDA) and solar energy technologies programme (SETP). SDA is especially relevant for efficient resources allocation. SAM in broad-based terms considers different types of financing and a variety of technology-specific cost models for almost the full range of available technologies. The SETP technologies at present represented in SAM mainly include the concentrated solar power (CSP) parabolic trough and dish-sterling systems in addition to PV flat plate and concentrating technologies. Finally, the levelized cost of energy in SAM software is determined in taking into account the sum of total installed cost and the sum of direct and indirect costs.

6.2.3.2 SMA Off-Grid Configurator

This particular software involves designing of plans and dimensioning of off-grid systems. The new type of SMA off-grid configurator from SMA offers a professional and customized solution for the simulation and dimensioning of an off-grid PV system. A unique feature of this software is that every design aspect is mapped out ranging from the dimensioning of the PV plant to the battery and inverter. The significant objective is to determine the profitability of the PV system and also the expected lifespan of the battery.

6.2.3.3 SolarPro

The solar system is exposed to the outdoor environment for the full day–night cycle. This software helps to calculate the amount of electricity produced taking into account the values of latitude, longitude and, importantly, the weather conditions of the installation site. This produces accurate simulation results. Importantly, simulation including the shadow influence due to surrounding buildings and objects allows the users to check optimal settings and module designs prior to undertaking the system installation. In addition, the software generates a current–voltage characteristic curve of the solar module precisely and quickly. This is essentially based on the electrical characteristics of each manufacturer's product. In total, shade, current–voltage curve, power and financial analyses constitute the four main functions of this software. It is well suited for almost all types of simulations with regard to solar electric power systems. The significant objectives of this software are as follows:

- Figuring out solar power over PV arrays
- Figuring out various shadows on arrays in an accurate manner
- Defining inverters, analyzing and creating graphs and reports
- Easy interface with animation features

6.2.3.4 PV-Design-Pro

This Windows-compatible software is designed to simulate the operation of PV energy systems on an hourly basis for a period of 1 year. It is mainly based on the climate and system design chosen by the user. Three versions of PV-DesignPro program are available and are included on the Solar Design Studio CD-ROM:

- PV-DesignPro-S for stand-alone systems with battery storage
- PV-DesignPro-G for grid-connected systems with no battery storage
- PV-DesignPro-P for solar water-pumping systems

Following are some significant outputs that are obtained by this innovative simulation software:

- Availability of solar fraction charts of the year on a monthly basis
- Battery states of charge on a monthly basis
- Performance table on a yearly basis
- Annual energy cost analysis based on costs of purchased energy and sold PV energy

In total, the charts cover every hour of the year and include solar radiation availability on a horizontal surface as well. Importantly, the software also gives an estimate of the rate of return which is normally worked out in terms of overall price per kWh of the system besides the payback years.

6.2.3.5 HELIOS 3D

Utility-based PV power plants are gaining footholds by the day across the globe. This software is a professional planning tool especially for the utility-scale PV power plants. This software permits shadow-free placement of the PV modules on a digital terrain at any geographical position and at any given date or time. Additionally, the software is capable of supporting all the phases of the project process. Some specific functions of this software are

Project development: This offers a meaningful analysis and evaluation of the terrain and yield rate.
Project layout: This involves structuring of the terrain, positioning of the mounting structures and optimizing the positioning to derive the maximum yield.
Project engineering: This deals essentially with electrical layout, bill of material along with list of GPS coordinates and cable lists and so on.

6.2.3.6 Helioscope

This software has been specially developed for grid-connected PV systems. It is utilized in a fully programmable mode with PV capacities of up to 5 MW. Also Helioscope has a huge component library with a facility of SketchUp shading integration. It is also linked to the meteorological database of the NREL. The software's design-integrated approach models a PV array based on its physical design. It thus leads to advanced modelling of system effects and enables powerful design and scenario analysis. To take care of advanced shade calculations, 3D obstruction models from SketchUp are directly imported into Helioscope. Thus, it gives an additive effect of SketchUp for 3D modelling with the simulation capabilities of Helioscope software.

6.2.3.7 HOMER 2

HOMER stands for hybrid optimization model for electric renewables. This is a well-recognized computer model which makes easy the activity of evaluating the design options for both the off-grid and grid-connected power systems. Key application uses pertain to stand-alone and distributed generation. It was originally designed at NREL for village power programme. HOMER's optimization together with the sensitivity analysis algorithms allows the user to work out the economic and technical feasibility of a large number of technology options. The allied concern is to take care of any uncertainty of the following few parameters:

- Cost of the technology(ies) involved
- Availability of the energy resources
- Influence of any other variables

The Centre for Energy Efficiency and Renewable Energy (CEERE) offers technological and economic solutions to environmental issues. Such issues arise from energy production, industrial production and commercial activities besides the prevalent land-use practices. HOMER is a computer model which makes easy the task of evaluating design options for both off-grid and grid-connected power systems for remote, stand-alone and distributed generation applications. A unique feature of HOMER is that it models both the conventional and renewable energy technologies. In 2009, NREL granted a license to distribute and enhance HOMER to HOMER Energy (i.e. another version of software). HOMER Energy provides a highly visible commercial outlet for NREL's renewable energy simulation tools. It is with a prime objective of enhancing the user of HOMER by industry and decision makers.

In total, HOMER is capable of undertaking three basic types of analyses: sensitivity, optimization and simulation. The power sources that can be easily modelled range from solar PV to fuel cells.

6.2.3.8 SOLERGO

SOLERGO is a simulation software that well suits especially the PV grid-connected systems. It was developed by an Italian-based Electro Graphics Srl to determine the output of a solar PV system interfaced to the grid. The software has a database of inverter, irradiation and, importantly, that of solar modules. It helps in giving a layout of PV system besides making available the consumption analysis, cable sizing calculation and, importantly, electrical diagram. The software is also enabled with a feature determining the payback in addition to knowing the pollutants that may be emitted. As the inverter is a key working component, it should be configured based on the shading effects both in the vicinity and remote areas. A unique feature of SOLERGO software is its interoperability with Ampere Professional (basically an electrical network calculator to analyze the electrical network of the PV generator, cables, surge protection device [SPD], protection and inverter). Further, it is interfaced with selective software such as CADelet and EPlus to obtain full drawing of the system besides its construction details.

6.2.3.9 PV Design Solmetric

Solar PV rooftop systems are now being installed in increasing numbers than ever before. PV designer solmetric is specially developed to provide the following functions:

- Fast workouts of key determinants
- Drawing of a roof outline
- Incorporation of SunEye shade measurements at specific locations on the roof
- Facility of drag and drop modules, size strings, check inverter limits
- Calculation of AC energy production for customized design

A special feature of this software is its exhaustive global databases of modules, inverters and traditional weather base. The facility is available to work out with various design scenarios besides comparing their AC kWh outputs side by side to know an optimum design choice. Solmetric is specially customized for residential and small-capacity commercial systems.

6.3 Comparative Features of Various Simulation Software

Table 6.1 sums up the important details of major software being currently utilized across the globe for simulation studies of solar PV systems.

6.4 Manual and Simulated System Sizing Procedures

In India, around 19,000 villages still do not have access to electricity. However, a vast geographical area has been provided with electricity via the locally available grid since many years

Table 6.1 Most Commonly Used PV Simulation Software

S. No.	Name of the Software	Name of the Software Developer	Cost/Licence	Name of the Website
1.	Solar Advisor Model	National Renewable Energy Laboratory, United States	Free of cost	www.nrel.gov/analysis/sam/background.html
2.	HOMER	National Renewable Energy Laboratory, United States	Free of cost	www.nrel.gov/homer
3.	PV F-Chart	University of Wisconsin, Madison, United States	$400 for single user, $600 for educational site	www.fchart.com
4.	PVSyst	Institute of Environmental Sciences, University of Geneva, Switzerland	900 CHF for a single machine licence, 150 CHF for additional machines	www.pvsyst.com
5.	RETScreen	National Resources Canada	Free of cost	www.retscreen.net
6.	TRNSYS	University of Wisconsin, United States	$2100 for Educational use	http://www.sel.me.wisc.edu/tmsys
7.	INSEL	Insel Company, Germany	€1700 for full version; €85 full version for students	www.insel.eu

(Continued)

Table 6.1 (*Continued*) Most Commonly Used PV Simulation Software

S. No.	Name of the Software	Name of the Software Developer	Cost/Licence	Name of the Website
8.	SolarDesign Tool	Verdiseno, Inc., Santa Cruz, United States	Free version available and expert version available with a monthly fee	www.solardesigntool.com
9.	SolarPro	Laplace Systems Co. Ltd., Japan	$1900 for educational use	www.lapsys.c.jp/english
10.	PV*SOL Expert	Dr. Valentin Energie Software, Germany	€2456 for 20 licences for educational use	www.valentin.de/en/products/photovoltaics/12/pv-sol-expert
11.	PV-DesignPro-G	Maui Solar Energy Software Corporation, Haik, United States	$249 for solar design studio CD-ROM	www.mauisolarsoftware.com

Source: Software attributes consolidated from multiple sources including www.pvresources.com.

now. Simply put, an electrical grid acts as an interconnected network that delivers electricity from the suppliers to consumers at large. It mainly comprises generating stations which are capable of producing electrical power in tandem with the high-voltage transmission lines. Their purpose is to transport or carry power from distant sources to demand centres. The distribution lines that join the individual customers are power systems energized by solar modules connected to the utility grid.

Grid-connected PV power systems are composed of (a) modules, (b) solar inverters, (c) maximum power point tracker (MPPT), (d) power-conditioning units and the grid connection equipment. A unique feature of grid-connected systems is the absence of batteries. At appropriate working conditions, grid-connected PV systems supply excess power, beyond consumption by the connected load, to the utility grid.

6.4.1 *Importance of Manual and Simulated Procedures*

Solar MW-scale power plants are now being installed in increasing numbers in India. Prior to this, solar decentralized systems for various end-use applications especially lighting and water pumping outnumbered the grid power plants for many years. However, there is now a growing tilt to demonstrate the effectiveness of (a) rooftop PV systems and (b) PV farms, that is ground-mounted installations. Irrespective of their mounting arrangements, the vital issue is to comparatively evaluate the manual and simulated energy procedures. A wide range of simulation software are available in the market, but PVSYST software is preferably used. PVSYST is a software package for the study, sizing and effective analysis of complete systems. This deals with grid-connected, stand-alone, pumping and DC-grid PV systems, for example the public transport. It includes extensive meteorological readings and PV systems components databases as well as general solar energy tools.

Professionals like architects, engineers and researchers gain through this easy-to-use PVSYST simulation software. The manually derived readings are collected from the supervisory control and data acquisition (SCADA) system along with the graphical representation of several parametric considerations. The manual data collected are to be compared with the simulated results that are drawn.

6.4.1.1 Key Elemental Considerations Connected with Manual Procedure

Solar PV power plants whether of a small or large capacity can be sized manually. However, there is a need to factor in a large number of both the major and minor parametric considerations. The underlying rationale is to determine the influence of sizing on the working of a PV power plant from the site-specific characteristics. This is to be followed up by knowing the efficiencies of component/sub-components which are used in the process of DC power generation, DC to AC power conversion and the grid interfacing of the AC power to the locally available grid power. No less important is the role played by the correct installation of components, component assemblies and so on. Steps that have to be manually followed are as follows:

- Thoroughly assess the solar radiation data for a designated location based on different organizations such as the American Space Agency (NASA), Joint Research Commission of the European Commission and the United Nations Environment Programme (UNEP). These enable to work out the amount of electricity generated. In a majority of simulation software packages, in-built type solar radiation databases are available.
- Try to get a land-use map of the designated location which shows clearly the various sites suitable for solar power generation.

- Firm up different locations on the land-use map and upgrade these as and when needed.
- Try to locate various building roofs that can be utilized for setting up the project based on a minimum roof area.
- Identify the access route to the grid power and accompanying need for grid connection.
- Ascertain the dimensions of the lands or roofs of the few buildings chosen likely to be used for the specified purpose.
- Carry out a thorough assessment of the following parameters much in accordance with the desired power requirement:
  - Type of roof/land area
  - Area of roof
  - Orientation of the roof
  - Pitch/slope of the roof
  - Strength of the roof
  - Influence of shading on the roofs
- Seek information on solar PV aspects from different sources both at the national and international levels.
- Prepare suitable layouts of the system vis-à-vis each of the selected buildings or lands.

6.4.2 Cumulative Output

Solar PV power generates power in a clean, safe, reliable and silent mode throughout the year. Thus, energy produced from a PV power plant is laid out via the use of a SCADA system on a daily basis. Such readings are generally taken on an average basis. As such, the average annual energy output is determined by simply multiplying the average monthly energy output with the total number of months in a year. Likewise, the daily energy output is also worked out for different months in a year. Similarly, the daily energy output is also evaluated for various months. In this way, monthly energy output is worked out by a simple multiplication of the number of days of a month with the daily energy output. In short, the following parameters are represented for specified purpose:

- Months
- Average daily energy
- Output (MWh)
- Monthly energy output
- Average monthly
- Energy output (MWh)
- Average yearly energy output
- MWh

With the availability of these values, it becomes possible to draw a graphical representation of the energy and solar energy generated in MWh on a monthly basis. Some months in a year will have the maximum possible solar energy generated.

6.5 Determination of Peak Variation and Possible Plant Rating

The average peak solar radiation for a month is determined in units of MWh/m². Some assumptive figure of area required for installation of solar PV power plant is chosen. The average peak solar radiations are calculated for various months. This paves the way for determining the possible plant rating by multiplying the average peak solar radiation value with the available area.

The given factors can be taken care of by manual practices and procedures. The following section takes a close look at a step-wise treatment of the PV power plant simulation via use of a specific software as a case specific example, that is PVSYST.

6.5.1 *Simulating with PVSYST*

PVSYST as a software with a wide outreach of use in India enables to have a preliminary as well as post-evaluation test data for feasible power generation. The cumulative system performance and efficiency of each component of the plant are determined by entering the make and specifications of a specific plant design. Key parameter is to emphasize on the average energy output via comparison with the available software package – PVSYST. The following types of inputs are readily needed to arrive at meaningful estimations of the system size and annual yield from a grid-connected PV power plant.

6.5.1.1 *Solar Module*

Selection of materials for cell/module fabrication is of prime importance. Materials could range from crystalline silicon to thin films such as amorphous silicon, cadmium telluride or copper indium diselenide. A maximum solar module rating is preferred for a grid power plant. This results in a lower area related cost apart from the visible advantages of handling, wiring together with reduced civil work etc. Series and parallel interconnection of modules is to be undertaken based on the maximum allowable voltage of the whole system. In India, this voltage value has been fixed at 1000 V as per the stipulated guidelines of Central Electricity Authority (CEA) of the Ministry of Power [2].

6.5.1.2 *Power Conditioning Unit*

Solar power is a DC source of power which needs to be converted into useful AC power. The key system that converts DC into AC power is known as inverter or power conditioning unit. It is important to choose the individual rating of an inverter much in accordance with the cumulative capacity of a grid power plant. This helps in fixing the total number of inverters to be used in a PV facility. Take for example an assumptive power plant with a capacity of 10 MW, which could possibly need ten 1 MW inverters or twenty 500 kW inverters. Currently, container-type high-capacity inverters of 1 MW capacity and above are available. These have the advantage of a plug-and-play facility which eases and speeds up the installation.

6.5.1.3 Transformer

Several types of transformers which are included in the system to step up the available voltage levels are currently available. The total capacity of a PV power plant is an effective indicator of knowing the total number of transformers to be used. Transformers are available as either a liquid-cooled or an oil-cooled one with both technical and physical specifications.

For a simulated environment to take shape, the factors as listed in Table 6.2 have to be entered in various templates that emerge during the simulation process. As a case-specific example, PVSYST software is described here.

6.6 Factors Contributing to Losses in PV Systems

Shading loss: In general, three types of shading losses are considered in the PV energy yield model. These mainly include (a) horizon shading, (b) shading between the rows of solar modules and (c) near shading due to neighbouring trees and buildings.

Incident angle: The incidence effect (the common term is incidence angle modifier [IAM]) relates to decrease of irradiance actually reaching a solar cell.

Low irradiance: The solar to electric conversion efficiency of a solar module decreases at low light intensities.

Table 6.2   Key Attributes of PV Component–Specific PVSYST Simulation Software

Module	Technology type Module dimensions Total area of the array field Module efficiency Module capacity Number of modules per MW Total number of modules per plant Number of series connected modules Number of parallel connected modules Total number of parallel strings Total array voltage Total array current
Inverter	Type of unit Capacity rating Number of phases Efficiency Total number of individual units Power rating Input DC voltage Input DC current Output AC voltage
Transformer	Type of unit KVA rating Number of units Number of phases Frequency rating Primary voltage rating Secondary voltage rating Primary current rating Secondary current rating Efficiency Connection arrangement
Grid specifications	Number of phases Voltage rating Frequency
PV power plant specifications	Cumulative capacity Voltage output Current output

Module temperature: The characteristics of a solar module are determined at standard test conditions (STC) of 250°C. The efficiency of a module decreases for every degree of temperature rise above 250°C.

Soiling: Losses occur due to the presence of dust and bird droppings.

Module quality: The majority of solar modules do not meet exactly the manufacturer's nominal specifications. Modules are generally sold with a nominal peak power and a specific tolerance within which the actual power is guaranteed to lie.

Module mismatch: Losses due to mismatch are mainly due to the fact that all the modules in the PV array do not have the same current and voltage profiles and there is a statistical variation between them.

DC wire resistance: Electrical resistance between the power available at the module and at the terminals of a PV array gives rise to what is commonly known as the Ohmic loss (i.e. I2R).

Inverter performance: Inverters are devices which convert available DC power from the PV arrays into usable AC power with a certain specified maximum efficiency. Depending on the inverter load, they will not always operate at the maximum efficiency.

AC losses: This generally includes transformer performance (MV/HV) and Ohmic losses in the cable due to substation.

Downtime: This mainly depends on the grid power availability, diagnostic response time, availability of spares equipment and repair response time.

Degradation: The performance of a solar module normally decreases with time.

Maximum power point tracking: The inverters constantly seek the maximum power point of the array by shifting inverter voltage to the maximum power point voltage. Different types of inverters do this with varying efficiency.

A few loss-making factors are explained in detail for clear understanding.

Reflection losses: Solar modules are rated at standard test conditions. These need perpendicular incident sunlight. Under the actual field operating conditions, larger incidence angles occur. This results in higher reflection losses than accounted for in the nominal power rating. As per the calculations solar modules faced towards the equator and with a tilt angle equal to latitude, annual reflection losses relative to STC are about 1%.

Soiling: The soiling of solar modules occurs as a consequence of dust and dirt accumulation. In many cases, the dirt materials are washed off the panel surface by rainfall. However, dirt like bird droppings may stay even after heavy rains. The most critical part of a solar module is the lower edge. Especially with rather low inclinations, soiling may take place at the edge of a frame. Dirt also accumulates when water gets collected in the shadow puddle between the frame and glass. Once it causes shading of the cells, this dirt reduces the power available from a module. The losses are generally 1%. However, the power is restored if the modules are cleaned.

Mismatch effects: Mismatch losses are caused by the interconnection of solar modules in series and parallel, which do not have identical properties or which experience different conditions from one another. Mismatch losses take place in the modules and arrays because the output of the entire PV array under the worst case conditions is determined by the solar module with the lowest output. Therefore, the selection of modules becomes important, which contribute to the overall performance of the plant.

Maximum power point tracker loss: The power output of a solar module changes with changes in the direction of the sun, changes in the solar insolation level and with varying temperature. In the power versus voltage curve of a module, there is a single maxima of power, that is there exists a peak power corresponding to a particular voltage and current. However, as the module efficiency is low, it is desirable to operate the module at the peak power so that maximum power can be delivered to the load under varying temperature and insolation conditions. Hence, maximization of the power improves the utilization of a solar module. MPPT is used for extracting maximum power from the solar module and transferring that power to the load. A DC/DC converter (step up/step down) serves the purpose of transferring maximum power from a module to the load.

Losses that occur in cabling, transformer, inverter and transmission systems are easy to determine in most cases.

Critical factors: Critical factors that should be considered during the design process are as follows:

■ Proper selection of the modules
■ Optimum angle of tilt
■ Minimization of ohmic losses via proper selection of conductors
■ Selection of efficient transformers
■ Selection of efficient inverters
■ Utilization of reliable components with a long lifespan

Effect of sunlight: Some of the received light is reflected from the surface of the modules and never reaches the actual PV material. The amount of light reflected depends on the angle at which the light strikes the module. The more the light comes from the side (narrow angle with the module plane), the higher the percentage of the reflected light. This effect does not vary strongly between the module types.

Effect of diffuse sunlight: The solar to electric conversion efficiency depends on the spectrum of solar radiation. Nearly all the PV technologies show good performance for the visible light. However, there are large differences for near-infrared radiation. If the spectrum of the light were always the same, this effect would have been assumed to be a part of the nominal efficiency of the modules. However, the spectrum changes with the time of the day and year and with the amount of solar radiation availability.

References

1. www.pvresources.com.
2. Technical Guidelines for Voltage Regulation: Central Electricity Authority (CEA), Ministry of Power, Government of India, www.cea.nic.in.

Chapter 7

Real-Time Application of PV Simulation Software

7.1 Introduction

Solar photovoltaic (PV) modules are exposed to outdoor environment throughout the year. This makes it all the more important to evaluate the performance of PV systems under actual field operating conditions. It is quite true for installations that exist either in the developed or in the developing countries, for example in Germany. Meteocontrol is an accredited German-based company, which undertook a massive exercise of reviewing the field performance of about 30,000 installations across the European region and found a very startling observation that around 80% of the systems were underperforming. This was also reported in the *New York Times* dated 28 May 2013. The era of PV grid power generation started in India in 2010 under the ambit of Jawaharlal Nehru National Solar Mission (JNNSM). This shows that Indian MW scale programme is still in its nascent stages and thus relatively new. Little information is available on such types of systems. However, a few organizations – Resolve Consultants, Chennai, and the one commissioned by the Central Electricity

Regulatory Commission (CERC) in 2012 – have studied the field performance of MW-capacity PV-grid power systems. The following section gives a brief insight into the key observations as recorded from these select few organizational pursuits through real-time monitoring of the PV power plants from various key elemental considerations.

7.2 Performance Ratio versus CUF

Performance ratio (PR) and capacity utilization factor (CUF) are two ways to approach the efficiency of a PV system. PR is expressed in percentage and describes the relationship between the real and the theoretical possible energy output of a solar PV plant. It shows the proportion of the energy that is actually available for export to the grid after deduction of energy losses (e.g. thermal losses and soiling) and energy consumption for operation. The closer the PR value of a PV plant approaches 100%, the more efficient the respective PV plant is operating. As losses are always expected, a PR of 100% cannot be achieved. In other words, if the calculated PR is greater than 100%, then there is a measurement error. PR gives insights as to how efficiently the available solar energy is converted into electrical energy. It is possible to compare the performance of various plants at different locations on a normalized – independent from climate conditions – level over a long period of time. Deviations in the PR (e.g. values below the expected range) indicate a possible fault or problem in the solar PV plant. PR can, therefore, be understood as an early warning system.

7.2.1 Levelized Cost of Generation

The levelized cost of generation is a simple ratio of total life cycle cost and total lifetime energy production, that is the total number of units generated during the expected lifespan of a PV power plant. A simple equation which relates the capital cost (i.e. CAPEX), running cost (i.e. the operational expenditure) and residual value and denominator of total lifetime energy generation is

$$LCOE = \frac{CAPEX + OPEX - Residual\ value}{Total\ Lifetime\ Energy\ Generation}$$

The capital cost (i.e. CAPEX) is a one-time cost which is normally spent on the procurement of power-generating equipment and its associated installation, whereas OPEX is a running cost which is generally in the range of Rs. 10–11 lacs per MW on a yearly basis. The residual value being a one-time cost is typically equivalent to 10% of the CAPEX. This is followed up by an all-important value of lifetime energy yield, which basically aims at the optimization of the plant performance. Obvious enough is the fact that a higher value of lifetime generation ensures a lower value of levelized electricity cost. In the existing scenario, the value of capital cost (i.e. CAPEX) is maintaining its downward cost trend. It is of an absorbing interest to list down two vital parameters: one is the asset creation base which is mainly carried out during the first year and the other is asset management over the full lifetime of a PV project. It is logical to think that everything from project conceptualization to commissioning exerts some influence on the generation in ultimate terms. Table 7.1 presents the individual elements related to asset creation and asset management in simple terms.

Routine operation and maintenance of a PV power plant are of utmost importance. Sufficient care is required to maintain spares in a good enough quantity at the site apart from the deployment of the requisite number of O&M personnel. It is also of vital importance to have a master technician at the site who is adept in handling any emergency situations as and when they arise. No less important is the requirement of regular cleaning of the vast PV array field at least twice a month. In total, maintenance requirements can be generally categorized as follows.

Table 7.1   Asset Creation and Management of Key Elements

Asset build-up phase	*Project Development*

	Proper siting of the available land area
	On-site assessment of incident solar irradiation
	Conceptual design and engineering
	Project financing
	Evaluation of the available range of technologies
	Selection of the EPC company
	    Construction, engineering, optimized design
	    Shading analysis
	    Spacing in between the rows
	    Procurement-component selection
	    Construction-build quality

Asset management	Typically known as the operation and maintenance phase of a power plant facility Grid availability Uptime of the grid system Pilferage Real-time performance of the system

7.2.2 Preventive Maintenance

This usually deals with the routine inspection and servicing of the equipment. A significant objective is to stop any breakdowns so that the loss of energy yield loss is kept at a reduced level.

Condition-based monitoring: This essentially involves routine monitoring of the equipment condition and plant operation on a real-time basis. The significant objective of doing so is to overcome any problem at the initial stage itself so as to prevent the plant downtime.

Corrective maintenance: This solely relates to the routine repair of broken equipments on a regular basis. It is useful to become aware about the various steps involved with the first of the maintenance types, i.e. preventive maintenance mentioned above. Table 7.2 lists out the preventive maintenance schedule of a PV power plant facility.

Table 7.2 Enunciation of Major Steps for Preventive Maintenance of a PV Power Plant

Step No.	Type of Maintenance Requirement	Brief Remarks
1.	Cleaning of the solar module	It is very important to wipe clean the glass surface of solar modules at least twice a week so as to remove any dirt/dust.
2.	Growth of vegetation	It is common to see vegetation growth after rainy season. It needs to be pruned to a desirable size.
3.	Testing of DC and AC sub-system	It is very important to carry out routine testing of the sub-systems both on the DC and AC side.
4.	Inspection of the mechanical side	It should be carried out to detect any kind of fault occurrence.
5.	Current–voltage testing	Current–voltage characteristics curve of a solar module is akin to the ECG test on a human body. It reveals information on the power-related parameters of a module.
6.	Calibration of the sensor	It is of an increasing significance to keep the sensor in a calibrated position.
7.	Security system	It is an important requirement to have a properly functioning alarm system in case of any fire accident due to a short circuit or any other malfunctioning of the electrical system.
8.	Entering the information	Several types of information modules can be prepared on a daily/weekly/monthly basis. It becomes important to meet the full range of documentation requirements both for ready reference and mitigation.

(Continued)

Table 7.2 (*Continued*)   Enunciation of Major Steps for Preventive Maintenance of a PV Power Plant

Step No.	Type of Maintenance Requirement	Brief Remarks
9.	Management of warranties	Several key components installed at the plant carry different types of warranty schedules/requirements. Several criteria need to be adhered to both on the plant operators side and the equipment sellers side.
10.	Fault diagnosis	Solar PV plant despite involving no moving parts can pose one problem or another at times. Thus, it is important to undertake a routine fault detection check as a preventive measure on the site.

Next to the plant maintenance requirements comes condition-based monitoring. Table 7.3 highlights the steps required for the proper functioning of a PV power plant facility during the course of its functioning.

Corrective maintenance primarily involves three key approaches: (a) emergency maintenance, (b) automated owner and technical plant alert using email and SMS facility and (c) failure analysis and generation-specific report. It is of vital significance to make a realistic analysis of the underlying problems which may be affecting the plant performance in one way or another.

7.3 CERC Field Assessment Study of PV Power Plants

CERC [1] commissioned a field assessment study covering about 23 solar PV power plant sites at various locations in the

Table 7.3 Functional Monitoring of a PV Power Plant

S. No.	Type of Maintenance Requirement	Brief Remarks
1.	Regular monitoring of inverter	It is regarded as the heart of a PV power plant and should be taken care of properly much like a human heart. So it is absolutely necessary to monitor its daily operation.
2.	Routine management of the energy metre	Energy metre is akin to our electricity metre but different in the sense that it records the number of units generated and not consumed. Thus, it is an important link between the actual working of the PV plant and its capacity to perform well.
3.	Real-time monitoring and verification	The modern-day technology makes it possible to undertake the real-time monitoring of an operational facility; in this case a solar PV power plant.
4.	Generation of the daily, weekly and monthly reports	Solar PV power generation is much in accordance with the climatic characteristics prevailing at a given site on a daily/weekly/ monthly basis. Obviously, generation is not expected to be the same but it needs to be recorded without fail every day.
5.	Performance ratio and generation report	Both these parametric considerations are of significant importance.
6.	String-level monitoring	A large number of strings are physically present as a result of series and parallel combinations of solar modules.

(Continued)

Table 7.3 (*Continued*)   Functional Monitoring of PV Power Plant

S. No.	Type of Maintenance Requirement	Brief Remarks
		Their monitoring on a regular basis reveals information about the actual working of solar modules from several key considerations.
7.	Manage anomaly alerts and alarm management	It is quite important to manage any variance in the programmable alerts as well as in the alarm system.
8.	Web portal	Suitable display of relevant information.
9.	Historical data trends	Solar PV power plant starts producing valuable power almost instantly when commissioned. So, data availability and its consolidation is an important area of activity and has to be done till the end of useful life of a power plant.
10.	Performance optimization	For the PV plant performance to be optimized, it is significant to keep a close tab on key components that can be considered for the plant optimization purpose.
11.	In-depth analysis report	True performing synergy amongst the various system components is required.
12.	Accelerated PV degradation report	Crystalline silicon modules degrade relatively less than their rival thin film technologies.

(Continued)

Table 7.3 (*Continued*) Functional Monitoring of PV Power Plant

S. No.	Type of Maintenance Requirement	Brief Remarks
13.	Security system (fire alarm, intrusion alarm)	Mock drills/activities may be arranged with a twin purpose of testing the readiness/ effectiveness of the complete sensing equipment for any fire or unwanted entry of strangers.

country. Accordingly, solar radiation data for these locations were taken into account for calculating the CUFs. For this specific purpose, use of RETScreen, a simulation software, was made. The underlying assumptions and the results for both the crystalline silicon technology and the thin film technology are described in Table 7.4. The data for 52 (23 from the MNRE booklet + 29 others) locations have been prepared using RETScreen software and radiation data from the MNRE handbook on solar radiation. The assumptions made in RETScreen are shown in Tables 7.4 through 7.6.

Table 7.4 Assumptive Values for Crystalline Silicon Modules

Type of Modules	X-Si (Crystalline Silicon)	
Power capacity	10,000	kW
Manufacturer	Moser–Baer	
Model	MBPV-CAAP	
Efficiency	13.0%	
Nominal operating cell temperature	47°C	
Temperature coefficient	0.43%	%/°C
Solar collector area	7,692	m²
Control method	MPPT	
Miscellaneous losses	7.5%	

Table 7.5 Assumptive Values for a Solar Inverter

Efficiency	96.0	%
Capacity	1,000	kW
Miscellaneous losses	0.00	%

Table 7.6 Assumptive Values for Amorphous Silicon Modules

Type of Module	*Amorphous Silicon Module*	
Power capacity	10,000	kW
Manufacturer	Moser–Baer	
Model	MBTF Power Series	
Efficiency	6.0%	
Nominal operating cell temperature	47°C	
Temperature coefficient	0.20%	%/°C
Solar collector area	16,667	m^2
Control method	Maximum power point tracker	
Miscellaneous losses	7.5	%
Inverter		
Efficiency	96.0	%
Capacity	1,000	kW
Miscellaneous losses	0.00	%

7.3.1 Capacity Utilization Factors as Determined

The average solar radiation is expressed in kWh/m^2 and the electrical output in MWh. Table 7.7 shows the CFU at various locations in the country.

It is quite clear from the tabulated values of CFU that it changes based on both solar radiation level and air temperature.

Table 7.7 Capacity Utilization Factors in Various Geographical Regions in India

S. No.	City	Average Radiation	Ambient Temperature	Crystalline Output	CUF	Thin Film Output	CUF	Optimum Tilt
1.	Srinagar	4.10	13.6	13,377.97	15.27	1373.51	15.68	34.1
2.	Delhi	5.09	25.1	1,611.9	18.40	1708.4	19.50	28.6
3.	Jodhpur	5.52	26.1	1,732.40	19.78	1845.0	21.06	26.3
4.	Jaipur	5.52	26.1	1,741.10	19.88	1854.40	21.17	26.8
5.	Varanasi	4.88	25.1	1,521.90	17.37	1609.20	18.37	25.3
6.	Patna	4.83	25.3	1,509.80	17.24	1596.40	18.22	25.6
7.	Shillong	4.54	16.5	1,510.05	17.24	1556.50	17.77	25.6
8.	Ahmedabad	5.35	27.5	1,643.20	18.76	1753.80	20.02	23.1
9.	Bhopal	5.23	25.3	1,635.35	18.67	1734.89	19.80	23.3
10.	Ranchi	4.70	24.3	1,484.00	16.94	1562.46	17.84	23.4
11.	Kolkata	4.50	26.9	1,378.60	15.74	1458.30	16.65	22.5
12.	Bhavnagar	5.70	27.2	1,743.20	19.90	1863.80	21.28	21.8
13.	Nagpur	5.12	27.0	1,563.27	17.85	1662.80	18.98	21.1

(Continued)

Table 7.7 (Continued) Capacity Utilization Factors in Various Geographical Regions in India

S. No.	City	Average Radiation	Ambient Temperature	Crystalline Output	CUF	Thin Film Output	CUF	Optimum Tilt
14.	Mumbai	5.03	27.5	1,506.13	17.19	1601.85	18.29	19.1
15.	Pune	5.41	24.7	1,648.50	18.82	1745.40	19.92	18.50
16.	Hyderabad	5.67	26.7	1,706	19.47	1818.70	20.76	17.5
17.	Visakhapatnam	5.13	28.4	1,537.20	17.55	1638.90	18.71	17.7
18.	Panaji	5.50	27.4	1,645.87	18.79	1756.70	20.05	15.5
19.	Chennai	5.36	28.8	1,560.40	17.81	1667.60	19.04	13
20.	Bangalore	5.47	24.1	1,642.90	18.75	1736.10	19.82	13
21.	Port Blair	4.73	26.2	1,420.00	16.21	1500.27	17.13	11.7
22.	Minicoy	27.2	27.5	1,487.30	16.98	1577.50	18.01	8.3
23.	Thiruvananthapuram	5.41	27.3	1,581.3	18.05	1682.5	19.21	8.5
24.	Chandrapur	5.12	27.5	1,562.59	17.84	1664.87	19.01	20
25.	Pahalgam	4.70	0.00	1,703.90	19.45	1698.50	19.39	34
26.	Gangapur	4.97	25.0	1,569.60	17.92	1659.70	18.95	26.5

(Continued)

Table 7.7 (Continued)   Capacity Utilization Factors in Various Geographical Regions in India

S. No.	City	Average Radiation	Ambient Temperature	Crystalline Output	CUF	Thin Film Output	CUF	Optimum Tilt
27.	Ludhiana	5.23	22.6	17,608.10	19.50	1801.80	20.57	30.9
28.	Manali	4.59	−1.6	1,664.50	19.00	1650.20	18.84	32.3
29.	Dehradun	5.32	11.4	1,837.40	20.97	1884.20	21.51	30.3
30.	Churu	4.92	24.1	1,555.70	17.76	1641.50	18.74	28.3
31.	Jaisalmer	5.17	25.9	1,609.10	18.37	1708.40	19.50	26.9
32.	Allahabad	5.79	25.9	1,822.50	20.80	1943.90	22.19	25.5
33.	Darjeeling	4.80	9.0	1,641.00	18.73	1663.60	18.99	27.1
34.	Dibrugarh	3.92	17.1	1,320.58	15.08	1357.42	15.50	27.5
35.	Kota	5.08	25.4	1,592.70	18.18	1686.70	19.25	25.2
36.	Palanpur	5.15	26.6	1,594.80	18.21	1694.90	19.35	24.2
37.	Vadodara	5.29	27.5	1,621.60	18.51	1730.20	19.75	22.3
38.	Bhuvaneshswar	4.82	26.9	1,476.63	16.86	1566.03	17.88	20.3
39.	Ahmadnagar	5.17	25.6	1,582.70	18.07	1678.87	19.17	19.1

(Continued)

Table 7.7 (Continued) Capacity Utilization Factors in Various Geographical Regions in India

S. No.	City	Average Radiation	Ambient Temperature	Crystalline Output	CUF	Thin Film Output	CUF	Optimum Tilt
40.	Machilipatnam	4.95	28.0	1,479.50	16.89	1573.60	17.96	16.20
41.	Mangalore	5.08	27.3	15,113.06	17.27	1608.91	18.37	12.9
42.	Coimbatore	5.12	26.2	1,512.30	17.26	1601.90	18.29	11
43.	Dindigul	5.00	24.9	1,485.40	16.96	1566.20	17.88	10.4
44.	Amini	5.76	27.4	1,690.90	19.30	1690.90	19.30	11.1
45.	Jallandhur	5.39	20.4	1,766.80	20.17	1856.30	21.19	31.3
46.	Rae Bareli	5.05	24.9	1,594.80	18.21	1687.60	19.26	26.2
47.	Nadiad	5.60	28.16	1,630.60	18.61	1741.80	19.88	22.7
48.	Okha	6.11	26.1	1,895.30	21.64	2025.60	23.12	22.2
49.	Bhatinda	5.08	23.4	1,648.70	18.82	1740.40	19.87	30.2
50.	Dindigul	5.00	24.9	1,501.40	17.14	1583.10	19.87	10.4
51.	Siliguri	4.85	19.4	1,626.00	18.56	1693.90	19.34	26.7
52.	Ajmer	5.14	24.7	1,633.90	18.65	1728.30	19.73	19.73

7.3.2 Case Study of Chandrapur Grid-Connected PV Plant

High-capacity (MWp) PV power plants came into being with the advent of JNNSM in 2010. Since then a cumulative capacity of more than 4000 MW has been realized as against the abysmally low PV grid power capacity of 2.12 MW till 2009. Several plants of varying PV capacities are now in operation across the country. The author was interested to know the actual generation data from a plant with an immediate purpose to compare them with the design data. There is an underlying aspect involved with this desirable requirement of this specific chapter/section. Let us discuss two examples. One is a PV power site in Chandrapur located in the western state of Maharashtra. The design data of the developer agrees very well with the simulated design data. In fact, the actual performance exceeds the estimated generation. Likewise, another major PV power plant developer, the Azure power plant facility, has reported higher performance during the first months of working. Data for a few months is available from two other plants located in Kolar and Belgaum. Efforts were made to get the actual generation data from these plants and compare it with the design data. Data for only 1 year are available for Chandrapur plant. The actual performance exceeds the estimated generation. The efficiency of inverters is clearly reflected in the performance of the plants and is shown in Table 7.8.

7.3.2.1 Analysis of the Generation Data versus the Design Data

Chandrapur is located in the western state of Maharashtra and has the following few key parameters:

Latitude: 20.0°N
Latitude: 29.3°E
Elevation: 226 m from mean sea level

Table 7.8 Comparative Values of Designed and Actual Generation Figures

Month (of Year 2009)	Generation in MWh		
	Designed	Actual	Our Model
January	130	154	151.89
February	160	154	152.41
March	170	170	170.44
April	173	159	162.80
May	141	151	156.36
June	90	107	111.84
July	85	94	97.60
August	75	93	96.70
September	123	116	118.78
October	147	144	144.43
November	155	152	149.20
December	144	156	152.42
Total	1593	1650	1664.87
CUF	18.18	18.84	19.01

Table 7.9 highlights the key climatic characteristics of the designated site as per NASA data.

Key derivatives:

■ Solar insolation is highest in the month of April at 6.64 kWh/m²/day followed up closely by insolation values in May and March.

■ The month of May experiences the highest air temperature of around 33.6°C.

The monthly generation values until 31 December 2010 are given in Table 7.10 along with the reasons for the decreased values of plant load factor.

Table 7.9 Site-Specific Climatic Characteristics

Month	Air Temperature (°C)	Daily Solar Radiation Horizontal (kWh/m²/day)	Wind Speed (m/s)
January	23.2	4.80	2.5
February	26.0	5.65	2.80
March	30.0	6.23	2.7
April	31.6	6.64	2.9
May	33.6	6.51	2.9
June	30.0	4.76	3.0
July	27.5	3.91	3.0
August	26.9	3.77	2.9
September	27.1	4.60	2.3
October	26.4	5.02	2.2
November	24.6	4.91	2.5
December	22.6	4.66	2.5
Yearly Average	**27.5**	**5.12**	**2.70**

7.3.3 Case Study of a KPCL Grid-Connected PV Power Plant

This is a brief performance assessment result of a 3 MW PV power plant undertaken by the prestigious Indian Institute of Sciences (IISc), Bangalore [2]. Karnataka Power Corporation Limited (KPCL) has set up a 3 MW capacity grid-connected PV power plant near Yalesandra village in Kolar district of Karnataka. This plant is one of the 20 PV power plants of similar capacity in India as on 31 July 2015. It is located at a latitude of 12°53′ and a longitude of 78°9′, and it occupies an overall area of 10.3 acres. Further, the plant has three segments with each segment having an installed capacity of 1 MW. Each segment has four inverters

Table 7.10   Solar PV Power Plant–Specific Generation Figures and Associated PLF Values

Month (Year of 2010)	Generation (kWh)	PLF (%)	Actual Reason for Less PLF
April	32,800	4.55	Failure of 1250 kVA, 415/33 kV oil-filled transformer
May	73,620	9.89	Plant working with substitute 500 kVA 415/333 kV oil-filled transformer
June	10,860	14.80	Rainy season
July	96,550	12.9	Rainy season
August	1,05,890	14.2	Rainy season
September	1,00,390	13.9	Rainy season
October	1,14,770	15.4	Rainy season
November	1,05,660	14.675	Less solar insolation than expected and grid failure of around 970 min
December	1,12,570	15.13	Less solar insolation due to cloudy weather and a short grid failure of 51 min

with a capacity of 250 kW each with about 13,368 modules made of single crystal silicon modules. Twenty-four modules constitute 1 array with a total of 557 arrays. Solar modules are connected in such a manner that a voltage of 415 V is generated at the output of each inverter. This is further stepped up to 11 kV by a step-up transformer and connected to the existing 11 kV grid.

The cumulative electrical energy generated by this plant during 2010 was 3.34 million kWh. Of these, around 3.30 million units were fed to the grid. In this case, solar modules did not ill perform. However, a key component of the system, that is inverters, posed a major problem which in turn affected the power generation in ultimate terms. Further effect of temperature variation of the modules was studied both on a daily and yearly basis. IISc made use of daily datasets. The following factors could be attributed to reduced power generation:

- Improper functioning of all the four inverters
- Higher plant downtime
- Occurrence of more number of cloudy days than the usual

Out of the 357 days of operation, there were 75 days on which the grid was off for less than an hour and 57 days with grid-offs for more than 1 h. In total, grid was not available for about 201.4 h.

7.3.3.1 Temperature-Dependent Effects

A study by IISc observed that efficiency is more sensitive to the module temperature than the solar insolation. Although the solar insolation level is more in March compared to January, the efficiency during March is low. This could mainly be due to the increase in the daily average temperature of PV modules during that month. It is also observed that the efficiency of modules decreases and reaches the minimum during the peak hours. This is mainly because of an increase in the temperature of modules. This negatively impacts the efficiency more during that time.

In this case, the efficiency of modules decreases from 14.5% at 30°C to 11.5% at 55°C. Furthermore, the temperature of modules increases with the increase in solar insolation and reaches the maximum during the peak insolation hours. It thus leads to reduced conversion efficiency. Thus, the capacity of a plant to produce maximum power is retarded during the peak insolation hours, when the availability of incident solar radiation is the highest. This is mainly due to an increase in the module temperature.

7.4 Design versus Actual Data Variations for an Off-Grid PV Power Plant

Ladakh province of Jammu and Kashmir (J&K) state is a scenic landscape in the country. Durbuk is one of the six administrative blocks of Leh district in Ladakh and has a population of around 6000 people. This was the most backward area in the whole region. It forms the north-eastern part of the district. There was not even a remote option of taking grid to this area from the grid available in Leh. The villagers were getting electricity from a 250 kVA diesel generator which was being managed by the Power Development Department of the state government. Durbuk is bestowed with good solar insolation and it thus prompted the installation of a 100 kWp PV power plant. Table 7.11 sums up the brief technical specifications of the solar PV system.

7.4.1 Energy Generation from 4 × 25 kWp Units

Table 7.12 shows the cumulative energy generation from a 100 kWp PV power plants subdivided into four units of 25 kWp each as recorded by the project developer. These data are the performance data of about 17 months falling into 2 calendar years of utilization.

The total generation as available from these four units of equivalent PV power capacity is around 1,01,755 kWh [3]. This points to a massive shortfall of the generation due to various reasons. Simulated software–based calculations are not in a perfect synergy with the data actually realized at the designated site.

7.5 International Experiences

The first report of this nature is from a study previously undertaken by Electric Power Research Institute (EPRI) in July 2010. This survey included six utility companies, PV monitoring system providers, vertically integrated manufacturers in addition to solar energy service providers. Some examples are

Table 7.11 Key Technical Specifications of a 100 kWp PV Power Plant in Durbuk

Key System Component	Description
Module	
Module type	Based on polycrystalline silicon cell technology
Module rating	75 Wp
Total number of modules	1360
Number of modules per unit	340
Battery Bank	
Capacity of the cells	1000 Ah at C/10
Total number of battery cells	480
Number of cells per battery bank	120
Total number of battery banks	04
Charge Controller	
Total number of charge controllers	04
Capacity of each charge controller	25 kW
Inverter	
Total number of Inverters	04
Capacity of each inverter	25 kW
AC Feeder Panels	
Total number of outgoing panels	04

SunEdison, SunPower and Southern California Edison. The objectives of this broad-based survey are as follows:

■ Operation and maintenance practices prevalent at the respective sites
■ Analysis of the reasons for failure

Table 7.12 Energy Generation of a Cumulative Nature from Multiple Units of a PV Power Plant

Unit No.	Cumulative Energy Generated (kWh)
I	25,909
II	26,494
III	24,294
IV	25,053

- Financial provision for O&M measures
- Cleaning of the module surface
- Plant monitoring and information collection
- Warranty management
- Key lessons learnt

Figure 7.1 shows that the survey focused on a total of about 368 PV systems, out of which, the SunEdison-based system equated to about 117 MW of PV capacity. As mentioned, 20%

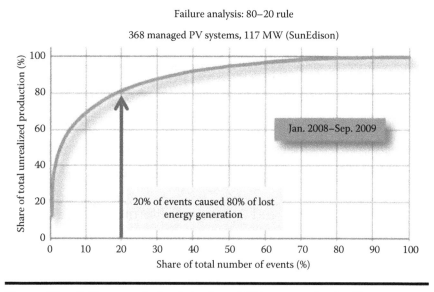

Figure 7.1 Failure analysis of deployed PV power plants. (From SunEdison, St. Peters, MO.)

Failure analysis—Failure areas

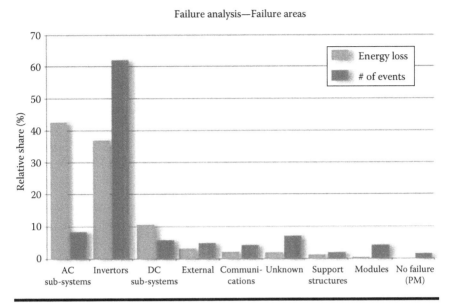

Figure 7.2 Typical problem areas in a PV power plant. Inverters failed the most number of times, but bigger impact was due to AC sub-systems. (From SunEdison, St. Peters, MO, 2009.)

of the events was responsible for a energy generation loss of about 80% [4].

Figure 7.2 highlights the important failure areas as a key part of failure analysis study. It was found that inverter was the reason for maximum number of failures, that is it failed the most number of times, but the larger impact was due to AC sub-system.

Figure 7.3 shows a graphical representation of the root causes of the failure analysis. The component use and construction quality impacts are the primary causes of energy loss in the PV systems.

7.6 Field Observational Study of Resolve Solar

Resolve Consultants is one of the most standard PV consulting companies in India. This company is well equipped on the consulting front and has served as a consultant to a large number

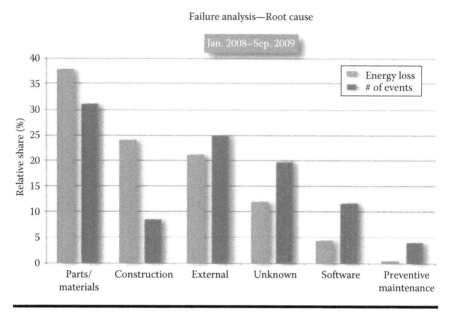

Failure analysis—Root cause

Figure 7.3   Key underlying reasons for leading energy loss in a PV power plant. Components used and construction quality impacts are leading causes of energy loss. (From SunEdison, St. Peters, MO, 2009.)

of project developers in India till now. Resolve Consultants has presented their on-site observations regarding PV power plants in MW scale installed across the diverse geographical regions of the country. The places that were included in the study were Rajasthan, Gujarat, Tamil Nadu and Andhra Pradesh amongst others. Importantly, PV power plants in all these places were commissioned after 2012. Though the capital cost of installing MW-scale power plants is high, their expected lifetime is between 25 and 30 years. Let us first deal with few financial terms which generally figure in the scheme of PV power generation or in any conventional power plant. Resolve Consultants which is a well-recognized PV companies in India was centred around some of the large-scale EPC companies/Developers such as:

■ Larsen and Toubro (EPC)
■ Vikram Solar (EPC)

- Waaree Energies (EPC)
- Sterling and Wilson (EPC)
- Tata Power Solar (Developer & EPC)
- Solairedirect India (Developer/IPP/EPC)
- Staten Solar (EPC)

The projects that were surveyed by Resolve had an aggregated capacity of around 560 MW and spread over the key states such as Rajasthan, Tamil Nadu, Maharashtra and Andhra Pradesh. The annual cost of operation and maintenance for majority of PV power plants was Rs. 10 lacs per megawatt. Figure 7.4 indicates the percentage contribution due to failure of the major components/sub-systems both on the DC and AC sides. Grid issues dominate the rest of failures, being responsible for about 82% of the fault occurrence. Inverters account for as much as 77% of the problems.

The mounting structures of solar modules/arrays cause few problems. Solar modules, despite so much of production

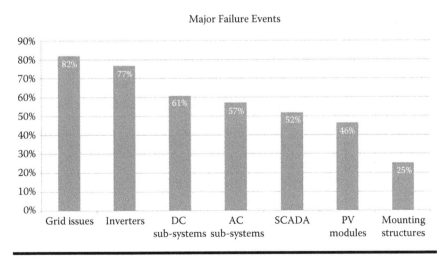

Figure 7.4 **Occurrence of component-wise problem in a PV power plant. Notes: (1) DC sub-systems: All components on the DC side before the inverter. (2) AC sub-systems: All components on the AC side of the inverter (including LV and HV).**

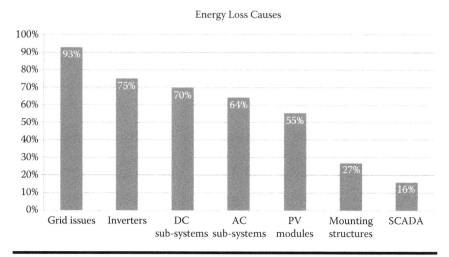

Figure 7.5   Major energy loss causes both in the DC and AC system. Notes: (1) DC sub-systems: All components on the DC side before the inverter. (2) AC sub-systems: All components on the AC side of the inverter (including LV and HV).

processes–related advancements, are still responsible for about 46% of the problems. Figure 7.5 highlights the major reasons for energy loss both in the DC and AC sub-systems. Grid issues contribute to about 93% of the energy loss compared to 55% loss via the solar modules.

In a PV power plant facility, several key components exist, effective combinations of which lead to DC and AC sub-systems. Figure 7.6 shows the important causes of failure in a PV power plant set-up.

Non-availability of a grid is responsible for about 66% of the failures, whereas the vandalism-specific failure rate is about 32%. Nearly 48% of the failure is due to poor construction.

7.6.1 Key Comments of the Respondents

■ The respondents feel that there should be a heightened focus on the performance ratio in place of absolute energy guarantee.

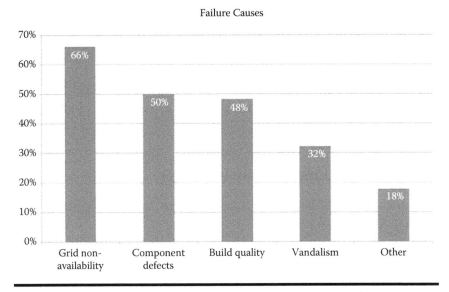

Figure 7.6 **Important reasons for failure in a PV power plant set-up.**

- ▪ It is absolutely important to carry out corrective measures from time to time.
- ▪ Capacity building measures need to be in place for the engineers and service technicians with still greater implementation focus.
- ▪ Major problem differentiators are visible in terms of wiping clean the glass surface of the solar modules and the grid power availability.
- ▪ Remaining losses happen to be very low. Losses in the inverter and solar management unit (SMU) seldom take place. Further energy losses due to AC system, structure and module are quite rare.
- ▪ The cost of maintenance and operation is high for PV power plant facilities in remote areas. As the solar radiation data available for such remote sites are generally poor, there exists a wide gap between the values of estimated generation and the actually realized generation.

7.6.2 Criticality of Major PV Plant Components

Table 7.13 shows a consolidated report of the problem areas in terms of criticality.

In order to overcome this, it is an important need to focus first and foremost towards the full range of electrical system.

7.6.2.1 Plant Optimization Measures

Following are a few suggestive measures to ensure smooth maintenance of PV system:

- Proper selection of solar module type/make/ratings is a must along with that of mounting structure.
- Appropriate components should be selected which are critical for the efficient designing of both the DC and AC sub-systems.

Table 7.13   Criticality of a Typical PV Grid–Connected Power Plant Operation

DC systems	Need definite attention although issues due to solar modules and support structure are not serious.
Inverter	Perceived to be the heart of a system, biggest internal controllable cause is in line with the internationally prevalent trends.
The grid	Turning out to be the major bottleneck. It has so far recorded the maximum level of failure events and thus accounted for the highest loss of generation. Incidentally, this is an external cause of a non-controllable nature.
Failure causes	The quality of both the construction and component use also figure amongst the recognized causes of failure.

■ Due focus should be on the selection of the key components/sub-components as far as possible as well as on the construction quality.

■ In case of both the DC and AC sub-systems, choice of the component is very critical. Specifically trained manpower is a vital requirement.

■ Prior to actual procurement, it is important to carry out an inside-out evaluation of the inverter(s). The following criteria must be met:

  – To ensure unfailing after-sales support from the source of procurement

  – To stock up the critical spare parts at the plant site itself

  – To enhance the diagnostic skills and capability of personnel as they are responsible for the routine operation and maintenance of the PV plant

  – To overcome the grid-related infrastructure problems

References

1. Central Electricity Regulatory Commission. www.cercind.gov.in.
2. A Case study of 3 MWp Scale Grid Connected PV Power Plant in Kolar, Karnataka, India. August 2011. www.dccs.iisc.ernet.in/3MWPV-Plant.pdf.
3. A 100 kWp PV Off-grid Power Plant in Durbuk, Ladakh, evaluation report by the Energy and Resources Institute (TERI), 2010 on behalf of Ladakh Ecological Development Group, Leh, Ladakh, Jammu and Kashmir.
4. Solar PV Operation and Maintenance: The Indian Experience, Presentation by Madhavan Nampoothiri, RESOLVE Consultants at *Inter Solar Conference and Exposition*, November 2013, Mumbai.

Chapter 8

Capacity-Building Initiatives for Simulation Software Outreach

8.1 Introduction

Solar PV technology has been on a roller coaster ride to penetrate both urban and rural areas globally. At one time, there was a wide subsidised market for end-use applications such as lighting, water pumping and battery charging in rural areas. As technology became more mature and reliable, markets began to shift towards urban areas. Rooftop PV systems came up in capacities ranging between just a few kilowatts and several megawatts. Importantly, PV grid–connected power plants gained ground in terms of a growing favour from a large number of project developers. For example, PV plants that had a capacity of just 2.11 MW have reached a capacity of 4000 MW within a short period of time, that is during 2011–2015.

It is important to point out here that at the outset, human resource availability was touted as one of the severe constraints. Taking a cue from it, several companies took up an early challenge to kick-start the human resource training programmes. These dealt at some length with the basic aspects of PV technology and then graduated to system design and engineering issues from several key considerations. Field visits to actual PV installations constituted an important activity of the training programmes. A selected few international organizations that dealt with capacity-building activities collaborated with some Indian companies to gain a competitive edge. The participation fee for such programmes, which last for 1–5 days, ranged from a few thousand rupees to around half a lac. Companies which came to be engaged in capacity-building initiative programmes became quite concerned to arrange the availability of various instruments. Compass, solmetric eye, hand-held pyranometer, portable multimeters and digital Ah readout metres found place in the training module kit of several PV companies.

This chapter traces the early history of capacity-building initiatives in India to the heightened focus on developing the skills and capabilities of technicians, engineers, managers and financial personnel from all possible considerations. As the programme expands and new PV technologies become available, the need for finding better trained resources will grow manifold. Currently, solar PV-specific materials such as wafers are not available in India. It can well be a part of the studies to think of ways and means to create such a capacity. In turn, interested personnel should conceptualize a broad-based approach and take it gradually to the stage of actual implementation on the ground.

8.2 Geographical Regions Generally Targeted

Solar energy is available in plenty and is present in almost every part of the world. In India around 400 million people do not have access to conventional electricity. Many developing countries in the Asian and African continents continue to have low per capita power consumption. In contrast, developed countries such as the United States have a higher per capita consumption.

'Solar' today makes better sense in terms of actual ground level implementation in the least developed countries. Nonetheless, the role of solar energy utilization is almost equally important in both the developing and developed regions of the world. This is because of technological innovations which generally take place in fast-emerging technologies like solar and help to make them cheaper. Declining costs of a technology like solar will pave way for substantial increase in volumes delivered. The outcome is a gradual shift towards achieving the much needed economies of scale.

Since the launch of solar programmes in the mid-1980s, India has successfully implemented solar energy. India has developed the entire range of PV technology cutting across the supply chain, which includes wafers, cells, modules and need-based systems. Based on the availability of these materials, a large number of products and systems were designed, developed and disseminated for meeting several end-use applications like lighting, water pumping and battery charging. Bangladesh has successfully deployed solar PV technology based on solar home system programme. Other countries like Indonesia, Senegal and Egypt are also interested in using PV technology.

The MENA programme aimed at the wholesome utilization of solar energy in the Middle East and African regions is being touted as a major initiative. Once implemented in full, it may see a healthy growth of PV technology/programme. All these case-specific examples have one thing in common – they seek initiatives to implement capacity building across various levels. The underlying rationale is to prepare a battery of dedicated and trained personnel for operation and maintenance requirements as well as for establishing successful entrepreneurship models.

Europe, especially Germany, has successfully demonstrated PV projects in several ways and manifestations. The solar roof-top system programme in Germany is still rated as a classical and highly successful programme worldwide. The 'Million Solar Roofs' programme began on a well-deliberated policy-cum-planning measure, and thus, 'feed-in-tariff' was coined as a widely used measure to encourage the utilization of PV technology. Geographically, solar scores well amongst the rest of renewable energy (RE) technologies. Today, the need for capacity building for technical considerations is as much important as training the stakeholders the ways and means to analyze a potential project idea from the key financial considerations.

It is quite true that even technologists with a wide exposure often fumble with financial analysis and vice versa. Thus, the need of the hour is to have efficient training sessions and education elements which are especially relevant to meet the requirements of both these categories in a smoothly implementable manner.

8.2.1 Training and Education Elements for Dissemination

For the sake of a wider programme outreach, India presents a suitable case of capacity building in the area of solar PV technology. It is pertinent to mention here that the Indian Ministry of New and Renewable Energy (MNRE) realized the strategic importance of training technicians and engineers to a high standard. The World Bank–supported projects on PV market development in India were implemented through the Indian Renewable Energy Development Agency (IREDA) way back in 1995. This multi-dimensional programme involved the premier academic and research establishments of India such as the Central Electronics Centre (CEC) at the Indian Institute of Technology (IIT), Chennai. Siemens Solar was incorporated into this programme for imparting the requisite education and training facilities at multiple levels of human intervention. Significant objectives of this major capacity-building initiative are as follows:

1. Training hundreds of technicians chosen from the industrial training institutes and polytechnic institutions of the country in terms of installation and maintenance practices.
2. Training the engineers chosen from the premier solar PV organizations for both classroom training and laboratory-based sessions. Field visits and on-site measurements of the key parameters were a special feature of this project.
3. Training the trainers with a specific aim of developing master trainers of high standard who could be deputed to wherever the need for capacity building arose.

The engineers were also given a first-hand exposure on the PV system simulation practices and procedures. Both the thumb rules and derived parametric checks were included in the simulation activity to create awareness in those entrusted with the key task of system design and engineering. The training sessions were interactive and actual demonstration of the commercially available products and systems was made. These systems were made available for the entire training programme by the key manufacturing organizations and system suppliers. In total, this market development project included all of the following ingredients:

1. Site resource assessment
2. Solar radiation availability analysis
3. Clear identification of appropriate end-use applications
4. System simulation activities
5. System design and engineering relevant to site-specific characteristics
6. Performance evaluation of the field-installed systems, that is spot checks of the installed PV systems
7. System diagnostics and troubleshooting
8. Information dissemination

8.2.2 Capacity-Building Initiatives Undertaken under the National PV Water Pumping Programme

MNRE embarked on a major programme of deploying around 50,000 solar water pumping systems in the capacity range of 200–3,000 Wp for the drinking water and related applications in the mid-1990s. This programme had the requisite support drawn from the state nodal agencies for RE, PV industry, financial intermediaries and NGOs. Capacity-building initiatives in this case ranged from basic system sizing to actual selection of the most well-suited system components much in accordance with the total dynamic head available at the site. This programme undertaken in a mission-driven mode helped to prepare a large human resource well trained in site resource assessment, understanding of the pumping requirements, system sizing and installation and commissioning. India, being a land of diverse geographical and climatic regions, is the best example for the utilization of (a) surface pumps, (b) submersible pumps and (c) floating pumps, minus any provision for energy storage, of course.

8.3 Third-Party Evaluation of PV Systems

MNRE did not bask in the sunshine of providing all possible financial and fiscal incentives to spur the growth of Indian PV industry. Additionally, it put into place a procedural mechanism of making an on-the-spot assessment of both the technical and socioeconomic impact generated by a wide range of products and systems, such as the following:

1. Solar lanterns
2. Solar home-lighting systems
3. Solar street-lighting systems
4. Solar water pumping systems
5. Solar community-based power plants

The author, for example, was actively involved in the wide range of technical discussions that took place between the team at the Energy and Resources Institute (TERI) and the concerned ministry/department from several key considerations. It offered an ample opportunity to fine tune the technical skills to devise suitable evaluation methodologies and procedures vis-à-vis the specified range of PV systems. About 17 PV power plants, also known as village power packs, located both in the hilly and plain regions of northern India were evaluated for around 4 years. Likewise, few organizations such as the Administrative Staff College of India (ASCI), National Productivity Council (NPC) and Indian Council of Agricultural Research (ICAR) assessed the performance of deployed range of products and systems under the actual field operating conditions. All these are categorized as capacity-building initiatives across the diverse range of geographical, demographical and entrepreneurship abilities.

8.3.1 Training Course for 'Solar Thermal and PV Field Engineers' for ITIs

A country-wide programme was organized by MNRE for around 6000 Industrial Training Institutes (ITIs). The course content was prepared by the TERI in coordination with the Directorate General of Employment and Training (DGET) under the Human Resources Development programme of the concerned ministry. This programme dealt with the classroom, laboratory and field-related work and involved the training of trainers as well. The author was the lead consultant of the research team, which engaged in wide-ranging discussions with the key stakeholders about the best possible ways to inculcate the requisite awareness amongst the technicians and engineers mainly. A large number of training sessions were organized at various venues across the country and fully supported by the host institutions. As a result, personnel were well trained and possessed the ability to handle operation and maintenance requirements (O&M of a variety of solar energy systems). The course content can be accessed online at the website of the concerned ministry (i.e. www.mnre.gov.in).

8.3.2 Capacity Building via National-Level Monitors

MNRE formulated a comprehensive plan of engaging the services of well-experienced and subject experts in its mission to evaluate the performance of field installed solar PV systems. They were designated as national-level monitors (NLM) whose responsibility was fixed in terms of making an on-the-spot assessment of various PV systems. Sufficient care was taken to assign such geographical areas to NLMs which were in close proximity to their areas of work. Thus, capacity building was created by the NLM who interacted with the caretakers of the PV systems located in the rural and semi-urban locations.

8.3.3 Capacity Building through Business Meets

IREDA is the financing arm of the MNRE. The primary role and responsibility of this financing institution is to make available soft loans to the existing/potential manufacturers/system integrators and the project developers. From its inception in 1987 till now, IREDA has been organizing business meets on diversified themes of the full spectrum of RE technologies at different venues throughout the country. Such forums are being utilized to augment the capacity-building initiatives already underway by different organizations and to make stakeholders well aware of the ongoing schemes/programmes of IREDA. Case studies regarding real-time installations form a necessary part of these business meetings which are normally filled to the capacity. Information dissemination in terms of leaflets/brochures/reports/magazines is also undertaken both in English and in regional languages.

8.3.4 *Capacity-Building Initiatives Based on Software Simulation*

Solar PV technology uses the available sunshine, thus implying its key importance in the overall dynamics of a solar PV system operation in an effective and safe manner. PV system simulation software have been used liberally since a long time now. As a case-specific example, the Indian Institute of Technology Delhi (IITD) has been engaged in imparting education and training to the users of the following simulation software.

8.3.4.1 *TRYSYS*

This specific software was generally viewed as cumbersome to use and involved several layers of clear understanding and practice sessions. It is pertinent to mention here that this software was made available by its foreign developer at a cost which was perceived to be somewhat cost prohibitive by the potential users in India. A special issue of TRYNSSY software is its utilization mainly for the solar thermal systems along with the building energy systems. This software was not much user-friendly and thus prompted interested users to find recourse in other simulation packages such as HOMER, NSOL, PVSYST and RETSCREEN.

8.3.4.2 *HOMER*

The National Renewable Energy Laboratory (NREL) of the U.S. Department of Energy (DoE) pioneered the development of 'HOMER' energy software both for the decentralized and hybrid system models. NREL has not widely involved itself in the task of capacity-building activities for an extended user base of HOMER software in India. The users have largely tutored themselves with the HOMER groups available worldwide and the case-specific examples of the software use available both on the HOMER website and elsewhere too.

8.3.4.3 PVSYST

Another simulation software in this series of designing well-engineered and well-performing PV systems is 'PVSYST'. This software has passed through a series of well-devised transformations with its extended feature basis. Trial versions with the available value-added features have attracted a large number of enthusiastic users across the country. However, no special training has been organized by the developer for the registered users of this software in India. Online support is readily available on the website of the software developer with the difference of having enhanced its cost right through its market inception.

PVSYST has attracted the attention of a large number of key stakeholders in India. With the advent of megawatt-scale PV power plants under the Jawaharlal Nehru National Solar Mission (JNNSM), this software has been well received. It is user-friendly and matches closely with the results obtained under field conditions. Financial institutions such as banks and non-banking finance companies seem to be favourably inclined towards the feasibility report/DPR generated by PVSYST.

8.3.4.4 RETScreen

It is now pertinent to mention here about yet another well-received simulation software – RETSCREEN is a clean energy-management software and is available free of cost. This is currently being used by more than 9000 professionals and students in India and its use is growing year by year. In fact, the use of this specialized software in the country spans government agencies at all levels, utilities besides institutions and private sector companies. One good example is the TERI where the author i.e. Suneel Deambi worked during 1990–1995. With due realization of the immense benefit possible via the use of RETSCREEN, TERI undertook an early initiative in 2004 to popularize its use at different levels. The institute has trained hundreds of students on the use of RETSCREEN to evaluate clean energy projects. This training included the highly specialized modules like 'Advanced Course on Sustainable Lighting Practices'. It was organized in association with Philips Lighting University and an online programme developed in collaboration with the Open University in the United Kingdom. Specific features of RETSCREEN as a training module according to Dr. Rajiv Seth, Registrar of TERI University, are as follows:

- A very useful tool for enhancing the analytical and decision-making skills of students.
- The five-step process of analyzing a project in terms of energy modelling, cost, greenhouse gas emissions, finance and sensitivity/risk leads to a quick and easy analysis besides a fast learning curve for the students.

In total, RETSCREEN is a software tool of very high value and in-depth analysis. It is currently available in a few Indian languages i.e. Hindi, Urdu, Bengali and Telugu.

8.4 Miscellaneous Initiatives at Capacity Building in India

The emergence of JNNSM in 2010 has changed the outlook of solar PV programme usability in India to a good extent. Prior to that, activities such as capacity building/programme outreach or information dissemination fit mainly into the ambit of government programmes mainly. The concerned ministry and state nodal agencies for RE used to play a pivotal role in popularizing PV programmes from this key perspective. Lately, several private organizations working in one way or the other on any of the key aspects of PV technology/programme implementation including its marketing/financing are offering their course curriculum of classroom and practical training types. As a result, a good number of technicians, system design and engineering personnel and experts from financing, marketing and social impact assessment areas have emerged. The following is a list of organizations which can be contacted for the specified purpose of imparting education, awareness and training on the diverse aspects of both the PV technologies and programmes/schemes:

- National Institute of Solar Energy, Gwal Pahari, Haryana
- National Centre for Photovoltaic Education and Research
- National Power Training Institute
- Energy Alternatives India, Chennai
- TERI, Delhi
- Swaminathan Research Foundation, Chennai
- Arbutus Consultants Pvt. Ltd., Pune
- IT Power India Ltd., Delhi
- Global Sustainable Energy Solutions (GSES), Delhi
- TUV Rheinland Laboratory, Bangalore
- Renewable Sunergy Services Pvt. Ltd., Delhi
- TRA International, Delhi
- Eazy Solar, Ahmedabad
- First Green Consulting, Delhi
- Steinbeis Centre for Technology Transfer India, New Delhi

The course content generally offered by these organizations ranges from site resource assessment on the one hand to system design and engineering and financing on the other. Participation fee per participant in these types of capacity-building programmes may be anywhere between Rs. 10,000 and 35,000 t. Recently, more organizations have started to host webinars that constitute the invaluable views of subject experts and people from industry at large. Few companies are also engaged in organizing interactive discussion sessions on topics such as solar policies in high potential states of India via internet. Normally, 50–60 participants are recruited free of cost and the registered participants are then encouraged to query/deliberate upon various discussion threads emerging during the discussions. Industry experts are normally invited to these sessions for the specific benefit of the participants at large.

8.4.1 Perspectives of Capacity-Building Initiatives

A large part of the socially oriented PV programmes is being implemented within the country by the state nodal agencies for RE. Often such agencies are occupied with the field implementation of the PV programme from several key considerations. This also includes the tendering, procurement and resource consolidation for their respective programmes. It is thus important to organize capacity-building programmes for their personnel on a regular basis. As these organizations also formulate various schemes and programmes for PV programme promotion, it becomes imperative for various stakeholders to share their overall experiences and future vision.

8.4.1.1 Impetus on Grid-Connected PV Power

The Indian Ministry of Coal, Power and RE Sources has targeted to achieve a capacity of 20,000–1,00,000 MW solar grid-connected power by 2022. This shows that there arises a huge demand for site resource assessment professionals, system designers, engineers, EPC players, O&M personnel, financial analyst, project managers and those taking care of programme documentation and outreach specialists. Due to this reason, a large number of education, training and awareness programmes especially on MW-scale PV grid power are being organized in more numbers than ever before. This has led to a sizeable demand of quality conscious and dedicated cadre of professionals, who are quite adept in imparting the desired skills and training amongst several layers of threshold interest to senior level capacity-building enrichment. Course curriculum that is easy to understand and of practical value and specially targeted at the grid power plants is being developed. Conferences and workshops held on the general theme of RE and sustainability devote a special session for the grid-connected PV power with an objective of increasing the standard of personnel such as engineers, technicians and outreach specialists.

8.4.1.1.1 Brief Summary of a Few PV Education and Training Centres in India

National Institute of Solar Energy

The National Institute of Solar Energy (formerly Solar Energy Centre) was established by MNRE in 1982. It is a unit of the MNRE, Government of India, dedicated to the development of solar energy technology and its related science and engineering. To achieve its objective, NISE has been working on various aspects of solar resource utilization and technology development in collaboration with other research institutions, implementing agencies and industry. Over the years, NISE has developed a variety of technical facilities for technology evaluation and validation, testing and standardization, performance reliability, monitoring and data analysis apart from training [1].

Location: Gurgaon, India (Campus)
Telefax: 0091-124-2579207
E-mail: sec.nic@in
Website: http://mnre.gov.in/centers/about-sec-2/

National Power Training Institute

National Power Training Institute (NPTI) is a National Apex body for training and human resources development in power sector with its corporate office in Faridabad. NPTI operates on an all India basis through its units in different power zones of the country located at Faridabad, Neyveli, Durgapur, Badarpur, New Delhi, Nagpur, Bangalore, Guwahati, Nangal and Centre for Advanced Management. NPTI has a vision of value orientation and value addition to power and energy sectors through training and human resources development, endeavouring to energize people who will energize the nation.

Location: Faridabad, India
Phone: 9818897666
E-mail: rkmishra@npti.in
Website: http://npti.in/default.aspx

The Energy and Resources Institute

The Energy and Resources Institute set up in the mid-1970s is one of the most well-recognized think tanks. It works on the full gamut of energy-environment solutions from a variety of end-use applications. The institute has done considerable work in the area of biotechnology and tissue culture not to mention its high-profile work in the area of climate changes. TERI is on the forefront of a large number of capacity-building initiatives across interdisciplinary areas. It not only enhances the concept of sustainability and livelihood generation but undertakes practical skill-based training programmes for the master trainers and trainers. Figure 8.1 [2] gives a quick glimpse of a PV education and training session in progress at the TERI premises.

Figure 8.1   PV education and training session in progress at TERI University.

Location: New Delhi, India
Phone: +91 1124682100
E-mail: info@teri.in
Website: http://www.teriin.org

Global Sustainable Energy Solutions

Global Sustainable Energy Solutions (GSES) is an RE engineering, training and consultancy company specializing in PV solar design, online and face-to-face solar training, publishing solar books and PV system audits. Established in 1998, the company has a diverse portfolio, executing projects in Australia, New Zealand, Asia, Africa and the Pacific Islands for both government and private enterprise. GSES is a leader in education and training in the Renewable Energy Innovation and Technology Sector. Figures 8.2 and 8.3 [3] show a pictorial representation of a PV training programme session being delivered by GSES, India, at ANERT, Kerala.

Figure 8.2   Field demonstration of a PV training session by GSES India.

Figure 8.3   A PV classroom session in progress.

It is also an active partner of government and private enter-prises and local communities on a global scale in facilitating the growth and development of the RE industry through edu-cation, training, engineering, consulting and publications.

Location: New Delhi, India
Phone: +91 1140587622
E-mail: info@gses.in
Website: http://www.gses.in/

Gujarat Energy Research and Management Institute

Gujarat Energy Research & Management Institute (GERMI) aims to develop human resource assets to cater to petroleum and allied energy sectors, improve knowledge base of policy makers and technologists and provide a competitive edge to leaders to compete in the global arena. In the past couple of decades, the country has become the fastest growing

Figure 8.4   Field view of a PV training session by GERMI, Gujarat.

discoverer of fossil energy resources. Figure 8.4 [4] gives a glimpse of a PV power plant–specific training session organized by GERMI at Gujarat.

India is also on the threshold of becoming one of the largest global markets. The need for a resource centre was anticipated to keep pace with the fast developing and competitive energy industry and to continuously build requisite intellectual capital and human resource capital.

Location: Gujarat, India
Phone: +91 079-66701362
E-mail: information@germi.org
Website: http://www.germi.org/
Facebook: https://www.facebook.com/GERMI.GRIIC

National Centre for Photovoltaic Research and Education

The National Centre for Photovoltaic Research and Education (NCPRE) at IIT, Bombay, was launched in 2010 and is a part of the Jawaharlal Nehru National Solar Mission of the Government of India. The objective of the centre is to be one of the leading PV research and education centres in the world within the next decade. NCPRE aims to create and execute the blueprint for human resource development for PV in India. The centre envisages basic and applied research activities, including silicon solar cell fabrication, energy storage, novel PV structures and development of power electronic interfaces for solar PV systems. The goal is to make solar PV a cost effective and relevant technology for meeting a significant part of the energy needs of India.

Location: Mumbai, India
Phone: +91 22-2576 4476; 44804479
E-mail: ncpre@iitb.ac.in
Website: http://www.ncpre.iitb.ac.in/

Barefoot College, Tilonia

The Barefoot College is located in the village Tilonia in Rajasthan. The college was established as early as in 1972 and is also better known as the Social Works and Research Centre (SWRC). The first major activity of the centre was the groundwater survey of the 110 villages of Silora block for the Rural Electrification Corporation. This project was completed in a record time of 24 months and resulted in the electrification of nearly all the villages in the block a decade later. The training and educating programme for personnel is of high standard and SWRC has good credibility amongst the large number of stakeholders. The institute has trained a large number of students in urban areas taking them to a higher level of achievement (Figure 8.5).

Location: Rajasthan, India
Phone: +91 01463 288351
E-mail: contact@barefootcollege.org
Website: http://www.barefootcollege.org

Gujarat Institute of Solar Energy

Gujarat Institute of Solar Energy (GISE) is an ISO 9001:2008 certified educational institute that provides technical training programmes on solar energy. The major objective of the institute is

Figure 8.5   A capacity-building session on solar PV in progress.

to produce human resources skilled in solar energy. The final goal is to promote the quality education so as to transform lives that will change the world for better future. GISE also wants to increase employability of unemployed youth by providing technical skill development training to one. The idea is to channel energies into solar energy usage and penetration in the remotest villages of India. It can then contribute to nation's growth with a much needed greener footprint.

Location: Gujarat, India
Phone: 079-2768 01 22
E-mail: info@gise.in
Website: http://www.gise.in/
Facebook: https://www.facebook.com/
    GujaratInstituteOfSolarEnergy

Arbutus Consultants Ltd.

Arbutus Consultants Pvt. Ltd. provides consulting and engineering services in the field of RE. It is one of the leading engineering consultants in the field of solar PV in India having been involved in MW-level grid-connected and smaller off-grid solar PV power projects right from the beginning of the national and state programmes. The company has provided consulting, advisory, design and engineering services for over 1000 MW of solar PV power since 2005. Arbutus has been on the forefront of organizing a large number of capacity-building programmes at various levels and across the diverse geographical regions of the country.

Location: Pune, India
Phone: 020-30484006/007
E-mail: arbutus99@gmail.com
Website: http://www.arbutus.co.in

Renewable Energy Centre MITHRADHAM

The Renewable Energy Centre MITHRADHAM is an Institutional Partner of the International Society for the Promotion of Environment and Renewable Energy ISPERE consisting of an informal group of committed experts. 'The first fully solar educational institution in India' is the major slogan. The members of the group are involved in generating good will from extended contacts opening up new possibilities, new ideas, concepts, approaches and new initiatives. MITHRADHAM is also a pilot NGO initiative in India for propagation of RE, a model for sustainable and holistic development, a model for organic cultivation of vegetables, fruits and spices and a model for Indo-German cooperation in environment and RE.

Location: Kerala, India
Phone: +91(0)484 2839185
E-mail: director@mithradham.org
Website: http://www.mithradham.org/live/welcome.php

Promoters and Researchers in Non-Conventional Energy – PRINCE

PRINCE (Promoters and Researchers in Non-Conventional Energy) is a volunteer-based group dedicated to creating a better world for everyone through the use of RE, through the promotion of biogas, vermicomposting, micro-hydro power plants, pedal power and other non-conventional energy sources. The group was started by Prof. Ajay Chandak who believes that protecting and enhancing the environment is possible only through using non-conventional energy sources. Besides other relevant activities, PRINCE provides to entrepreneurs free training in manufacturing solar devices, extended free consultation to all social organizations for use of non-conventional energy and education to people in energy conservation.

Location: Maharashtra, India
Phone: 91 2562 271795
E-mail: chandak@princeindia.org
Website: http://www.princeindia.org/

References

1. National Institute of Wind Energy, MNRE, News Items & Reports, 2015, http://nise.res.in.
2. The Energy and Resources Institute (TERI), New Delhi, 2015, www.teriin.org.
3. Global Sustainable Energy Solutions India, New Delhi, 2015, www.gses.in.
4. Gujarat Energy Research and Management Institute, Gujarat, India, 2015, www.germi.org.

Index

Milton Keynes UK
Ingram Content Group UK Ltd.
UKHW040444071024
449327UK00020B/967